Pra

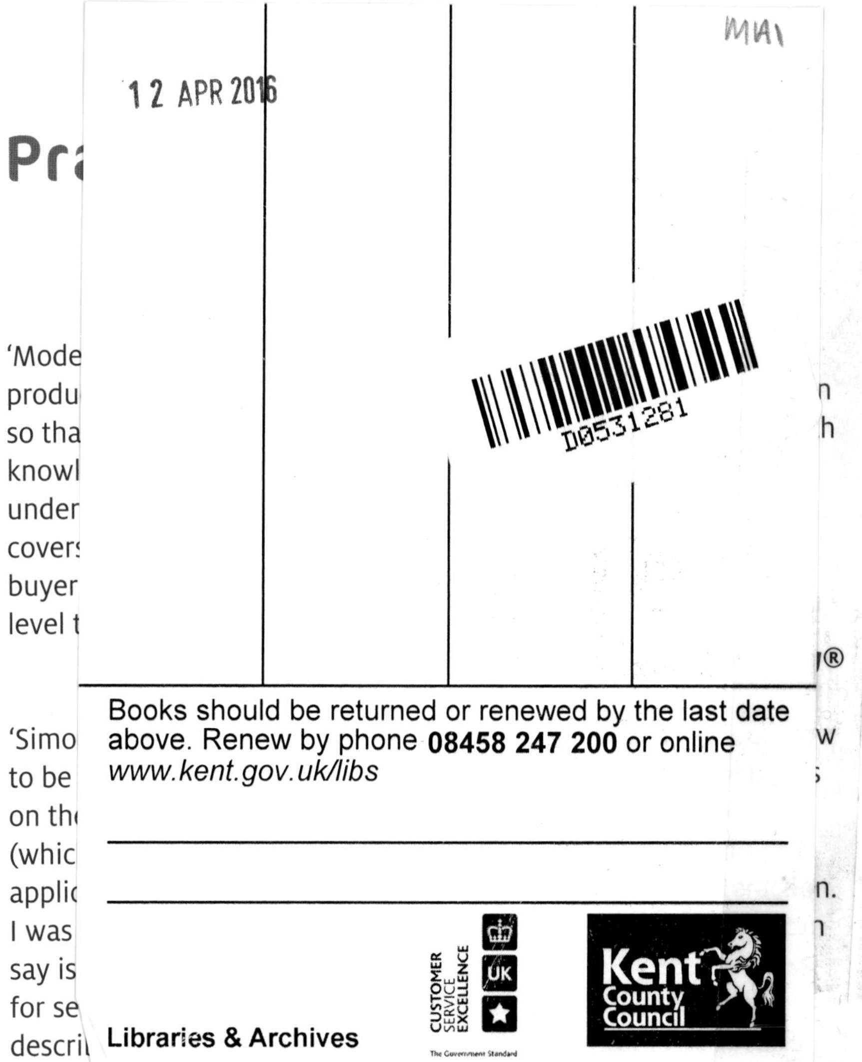

'Mode
produ
so tha
knowl
under
covers
buyer
level t

'Simo
to be
on th
(whic
appli
I was
say is
for se
descri

and demonstrates ninja-style effective techniques to get the results you need. I just hope my competition doesn't get hold of this masterpiece. It is a book to be studied, not just read. Get it. Devour it. Apply it. Your business and your life will be better for it.'

Terry Brock, MBA, CSP, CPAE, author, professional speaker, marketing coach

'Simon Hazeldine has conducted extensive research into neuroscience to understand how the brain responds during sales and negotiation and when a person is making buying decisions. His insights have created a highly effective sales tool that will help you close more sales with less effort.'

Allan Pease, author of the number one best-seller ***ody Language***

'Simon Hazeldine has taught me so much about how the brain works, and as a professional property investor it has helped me make a lot more money. I thought I could work people out until I read this book! Simon's neuro-sell, brain-friendly selling information is something you'll wish you knew years ago, and you'll realise just how much money you left on the table. Once you've read this though, it will ALL change. Do it now.'

Rob Moore, best-selling property investment author and co-founder of Progressive Property

'Simon Hazeldine's latest book *Neuro-Sell* is a must read if you want to know how to increase sales and understand the science behind why people buy. When you understand the science of neuro-selling you understand how to best adapt your approach and as a result you **will** win more business.

Simon successfully takes the complex issue of how our brains work and provides an easy-to-understand, practical road map to give anyone who reads his book a significant advantage in how to use the science of neuroscience to significantly increase sales. Great book Simon!'

David Tovey, business development speaker, author of *Principled Selling*

'Simon Hazeldine's *Neuro-Sell* combines the latest neuroscience research with hard-won sales experience to introduce the reader to the power of "brain-friendly selling". Essential reading if you want to create long-term client relationships in a world where technology has levelled the persuasion and influence playing field.'

Jamie Smart, author of *Clarity: Clear Mind, Better Performance, Bigger Results*

'*Neuro-Sell* is a fascinating and compelling read, which manages to get the right balance between science and practical application. It translates leading-edge research into simple, easy steps and actions that anyone can take to be a more effective salesperson. Anyone who needs to sell needs to read this book.'

Heather Townsend, author of *The Financial Times Guide to Business Networking*

NEURO-SELL

NEURO-SELL

How neuroscience can power your sales success

SIMON HAZELDINE

LONDON PHILADELPHIA NEW DELHI

Publisher's note

Every possible effort has been made to ensure that the information contained in this book is accurate at the time of going to press, and the publishers and author cannot accept responsibility for any errors or omissions, however caused. No responsibility for loss or damage occasioned to any person acting, or refraining from action, as a result of the material in this publication can be accepted by the editor, the publisher or the author.

First published in Great Britain and the United States in 2014 by Kogan Page Limited

2nd Floor, 45 Gee Street
London EC1V 3RS
United Kingdom
www.koganpage.com

1518 Walnut Street, Suite 1100
Philadelphia PA 19102
USA

4737/23 Ansari Road
Daryaganj
New Delhi 110002
India

ISBN 978 0 7494 6921 4
E-ISBN 978 0 7494 6922 1

British Library Cataloguing-in-Publication Data

A CIP record for this book is available from the British Library.

Library of Congress Cataloging-in-Publication Data

Hazeldine, Simon.
Neuro-sell : how neuroscience can power your sales success / Simon Hazeldine.
pages cm
ISBN 978-0-7494-6921-4 (pbk.) – ISBN 978-0-7494-6922-1 (ebook) 1. Selling.
2. Selling–Psychological aspects. 3. Neuromarketing. 4. Customer relations. I. Title.
HF5438.8.P75H39 2013
658.8501'9–dc23

2013032119

Typeset by Graphicraft Limited, Hong Kong
Printed and bound in Great Britain by CPI Group (UK) Ltd, Croydon, CR0 4YY

For KP and TJ as always

Contents

About the author

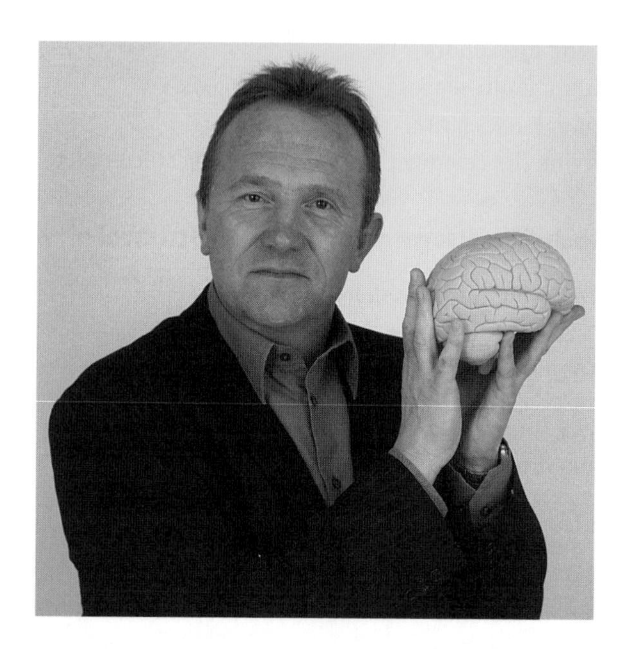

Simon Hazeldine works internationally as a professional speaker, performance consultant and corporate trainer in the areas of sales, negotiation, performance leadership and applied neuroscience. His focus is on inspiring and enabling exceptional performance and delivering improved bottom-line results for his clients.

Simon is the best-selling author of five books: *Bare Knuckle Selling*, *Bare Knuckle Negotiating*, *Bare Knuckle Customer Service*, *The Inner Winner* and *Neuro-Sell.*

He has a Master's degree in psychology; he is a Fellow of the Institute of Sales and Marketing Management, a long-standing member of the Professional Speaking Association and a licensed *PRISM* Brain Mapping practitioner.

Simon is also the co-founder of **www.sellciusonline.com** – the leading online resource for sales professionals.

Simon's client list includes Fortune 500 and FTSE 100 companies, and as a highly experienced and in-demand international speaker he has spoken in over 30 countries across six continents.

Foreword

In my roles as a board chairman and director of nine different companies, I know that to succeed in the modern world of business you need to do a better job of finding, getting and looking after your customers than your competitors do.

Although this may seem obvious, it is much easier to say than to do. The profession of selling needs to be transformed from one populated by far too many pushy, commission-hungry sales people to one populated by ethical, customer-centric sales professionals.

This book uses cutting-edge neuroscience research and insight to produce a highly effective, brain-friendly selling process that is powerful, ethical and customer friendly.

This groundbreaking book provides powerful cutting-edge insights that will give you an unfair advantage in selling situations. I particularly like the way that Simon takes what is clearly a very complex science and communicates it in easy-to-understand and practical terms that can be easily understood and applied.

Take the lessons from this book, apply them in your business and join the new world of brain-friendly selling. You'll be glad you did. And so will your customers.

Mike Greene
Star of Channel 4 TV programme The Secret Millionaire
Author of Failure Breeds Success *and* Into the Eye of the Storm
www.mikegreene.co.uk

Acknowledgements

My wife Karen, who in addition to providing countless hours of helpful proofreading is without a doubt the kindest and bravest person I know.

My son Tom for putting up with a dad who has spent far too many weekends locked away in his study writing this book.

Colin Wallace, PhD, formerly of the US-based Center for Applied Neuroscience, for all of his support, generosity and input.

My fellow co-founders of **www.sellciusonline.com**, Phil Jesson and Graham Jones. Thank you for your help and support and lots of chips.

Sue Richardson for connecting me with Kogan Page in the first place – you are a superstar!

Mike Speight for the superb photography.

And finally my editor Liz Gooster, who has helped to make this a far better book.

Introduction

It would appear that in our modern world it is not possible to turn on the television, read a newspaper or visit a news website without seeing some reference to neuroscience – the fields of science dealing with the structure and function of the nervous system and brain. Neuroscience has seen many of its major discoveries within the last 10 to 15 years, and thanks to powerful neuro-imaging technology new discoveries are being made on a regular basis. Neuroscientists are increasingly able to understand how our brains function and influence our behaviour. And as a result companies are increasingly looking to neuroscience for a commercial advantage. The list of companies currently involved in some form of neuroscience research is extensive. They want to understand how to interest people in their products and services and most importantly how to influence them to buy.

I am not a neuroscientist. I am an author, speaker and consultant who specializes in sales performance. Perhaps most importantly, I am an active sales professional myself. I have my own company, and if I don't sell successfully my profits suffer. Although I find the latest neuroscience research to be of interest, I am a pragmatist. My primary interest in neuroscience is in using it to improve my ability to sell successfully. The question I asked myself, which led to this book, was 'How can I use neuroscience research into how the brain functions so that I can then use this insight to enable me to sell to my customers (or more specifically sell to their brains!) more successfully?' The answer to that question, which is based on extensive research, lies within the pages of this book. Although in this book I quote the comments made by a number of eminent neuroscientists, I should make it clear that the overall interpretation I have given of the relevant neuroscience is my own.

The human brain is the seat of decision making, where decisions to take action and go ahead with a purchase or not are located. In this book you are going to learn how to work positively with the human brain that resides inside your customer's head. You will learn how to ensure your sales approach and methodology are 'brain-friendly', which will result in your customer's brain becoming increasingly receptive and welcoming of your sales message (and you!), thereby maximizing your chances of sales success. You will then have a distinct advantage over your competitors who understand less about how the brain works. They may have a very 'brain-unfriendly' approach that results in their customer's brain rejecting them and their sales offer at a very early stage in the sales process.

I think it is appropriate at this early stage of the book to discuss the concept of ethics. Undoubtedly there will be some people who are concerned that what neuroscience is discovering (for example about how people make decisions to buy certain products) may lead to the ability to manipulate people to take certain actions, perhaps even against their better will or judgement. Within the pages of this book you will learn some very powerful principles of persuasion that if correctly and intelligently applied will significantly enhance your ability to persuade others to take certain actions. My definition of 'correctly and intelligently applied' involves ensuring that you are adopting a customer-centric approach to your selling. This is based upon understanding what the customer needs and deciding if you are able to support the customer in achieving this aim. It is about helping customers to come to a decision that is right for them. If you do your job properly, then the majority of the time this will involve customers choosing your products and services. However, in some cases they would be better served by someone else, and the faster you discover this the faster you can move on to a more viable prospect for your products and/or services.

If you wish to succeed in sales in the medium to long term then building a base of satisfied customers whom you can use for referrals, references and testimonials should be a major priority. Hence my emphasis is on adopting a customer-centric sales approach.

The process outlined in this book will help the people you are selling to better understand what it is that they need, so that they can make better buying decisions. To some degree they will be persuading themselves! This can be achieved by:

- asking the right sort of questions that raise potential customers' awareness of their requirements;
- helping them to understand their decision-making processes;
- providing the specific information that they need at each stage of the decision-making process;

- delivering the information they need in a manner that suits their individual preferences.

When you sell in this manner you are maximizing the chances of the decision being a favourable one for you too.

Outside of the world of selling a great deal of interaction between human beings involves some form of persuasion. From persuading an employer to offer you a job or persuading a well-qualified candidate to join your company, to persuading your children to behave in a specific manner, to persuading someone you find attractive to go on a date with you or persuading someone to sponsor you for a charity fund-raising event, persuasion is a fundamental part of human behaviour. We are often engaged in the act of persuasion.

Anyone adopting a manipulative approach to selling will rapidly discover that this produces only very short-term results (you will get caught out at some stage) and a very bad reputation. And, in today's highly connected world (as a result of the rise of developments such as the internet and social media), your bad reputation may rapidly become apparent. It may also result in you not being comfortable with the face staring back at you from the bathroom mirror either.

Persuasion expert Dave Lakhani states in his book *Persuasion: The Art of Getting What You Want* (2005) that 'Your intention will ultimately determine whether you've persuaded or manipulated.' The techniques in this book could be abused, but that fact alone should not prevent them from being shared with sales professionals who will use them to help others and in doing so help themselves. The late Zig Ziglar, the legendary US motivational speaker, expressed it this way: 'You can get anything in life you want if you'll just help enough other people get what they want.'

Persuading people ethically and helping them to make decisions that truly benefit them will give you a solid foundation on which to build your future sales. The methodology outlined in this book will provide you with powerful persuasion principles that will power your sales success. Leaving people in a better place than when you met them will guarantee your long-term success in the sales profession.

1
The harsh reality facing sales professionals

Selling has always been a challenging profession. In sales your success, or lack thereof, is always obvious. Either you succeed in bringing in the business and hit your target or quota – or you don't.

It is easier to measure and manage the performance of people in sales than in any other profession. If you aren't making your numbers you feel the heat!

We live and work within a capitalist system. Competitive markets are one of the key components of capitalism. In a capitalist society competition in business is a fact of life, and history shows that competition tends to increase and get more intense. For those of us in the sales profession, an increasingly challenging commercial environment is the reality we have to live with.

And then there are other factors that are adding to the challenge that sales professionals face:

- Economic downturns have hit businesses hard. The effects of recessions last long after an economy starts to recover. Businesses that have become more financially cautious as a result of tough economic times do not always release the purse strings. Many learn to manage with less expenditure than they did previously and decide to keep it that way.

- The rise of globalization with the economic growth of countries such as Brazil, Russia, India and China has introduced new competition into Western markets.
- Products (and companies) are constantly being replaced by more effective or cheaper alternatives. Economist Joseph Schumpeter (1950) referred to this as a process of 'creative destruction' and described it as 'the essential fact of capitalism'.
- Product margins decrease as a result of increased competition, rising production costs and more aggressive and professional procurement practices. Over time these factors will erode the profitability of every product or innovation. An ongoing trend towards commoditization exists in many markets.
- The amount of buyer time allocated to sales professionals is declining (for example caused by downsizing of procurement departments), driving a need for sales professionals to be able to maximize the reduced face-to-face sales time they do manage to secure.
- An ongoing trend towards a more participative style of management has led to a greater empowerment of employees, with greater levels of personal accountability and a sharing of corporate goals. This helps to develop pride in expertise and emphasis on quality and doing a good job. This has led to buyers who take a strong personal interest in driving a hard bargain!
- The growing availability of information via internet-based research is increasingly leading to buyers who are better informed, and therefore possess greater levels of market knowledge or expertise than previously. An added complication is that as a result of their research many buyers are more likely to be misinformed (a little knowledge can be dangerous after all!), placing an additional demand on sales professionals to develop the ability to sensitively re-educate buyers.
- In order to maintain margins many industries are shifting their marketing and sales focus from product to services and solutions. This has created a demand for sales professionals who can make the transition from a transactional 'box-shifting' approach to a consultative 'value-added' approach. Many transactional salespeople are struggling to make this transition.
- Cost of sales is becoming a major concern for many organizations. Personal selling is the most expensive method of transferring goods and services from manufacturer to customer and according to

research is responsible for up to 55 per cent of total sales and marketing costs. Surveys show that some traditional sales methods (such as cold calling) are becoming less effective (and therefore more expensive) than they used to be.

As a result of the many challenges mentioned above, companies are striving to maximize the performance of their existing sales force, and sales professionals are feeling the pressure. Therefore sales professionals need an edge – the latest neuroscience research is that edge.

In this book you will learn about 'brain-friendly selling' that will make the whole process easier for everyone involved. We can now ensure that our sales messages are brain-friendly and appeal to the parts of the brain (both conscious and unconscious) involved in decision making.

Sales professionals are a vital component of a capitalist society – every sale 'is capitalism writ small' (Knight, 2008). Without sales professionals businesses would fail.

As I am fond of saying in my keynote speeches, 'In business nothing happens until someone sells something!' Selling is one of the most important jobs (if not the most important!) in the commercial world. My hope is that what you learn from this book will make that important job just a little easier and bring you greater levels of success and achievement.

I believe the future of your sales success, and indeed the future of selling, lies within the three pounds or so of brain cells contained inside your customer's head. The human brain is the most complex structure in the known universe, so let us begin by exploring it and understanding it.

2
The background to neuroscience and how it applies to selling

The human brain is widely acknowledged as the most complex, flexible, best-organized, highest-functioning system in the known universe.

Our brains control nearly everything we do. The brain appears to be a relatively small part of the human body – weighing about 3 pounds and making up about only 2 per cent of our body weight. However, it is estimated that the human brain contains 100 million nerve cells known as neurons and uses 20 per cent of the oxygen we breathe and 20 per cent of the energy we consume. The enormous consumption of oxygen and energy is required to fuel the many thousands of chemical reactions that take place in the brain every second. These chemical reactions underpin our actions and behaviours. This theory is, however, complicated by many facts: different levels of these chemicals can produce different effects. These substances do different things in different brain parts. Each interacts with others in different ways under different circumstances. Each harmonizes with many other bodily systems and brain circuits, setting up complex chain reactions.

It should be pointed out that neuroscience is the scientific study of not just the brain but the rest of the nervous system too. The brain is the central part of the nervous system. The nervous system allows us to respond to what we experience in the world around us.

The nervous system has two main divisions. The central nervous system is made up of the brain and the spinal cord. The nerve fibres that branch out into the body from the central nervous system form the peripheral nervous system. The peripheral nerves are constantly sending information to the central nervous system, which processes it and then sends signals back to the peripheral nervous system.

This book explores how we can apply knowledge gained from neuroscience research about how the brain and the rest of the nervous system operate to gain an advantage in persuading other people to make decisions that benefit themselves and us as sales professionals as a result.

For the sake of reading ease I shall refer to 'the brain' in its widest sense as an integral part of the wider nervous system. So unless specific reference is made to a specific part of the nervous system or brain you can assume I am referring to the wider nervous system.

For decades neuroscientists and neuropsychologists have been researching and studying the human brain to better understand how it functions and what brain processes influence our behaviours and actions. Knowledge and understanding about the brain are growing rapidly, with the vast majority of major discoveries and knowledge about the brain having been made in the last 10 to 15 years. More research has been conducted and more published on the brain in this time period than in the whole of human history.

One of the reasons for this acceleration of knowledge is the availability of brain-scanning technology such as electroencephalography (EEG), which uses sensors to capture the tiny electrical signals that brain activity produces. Other technology includes functional magnetic resonance imaging (fMRI), which measures the increase in oxygen levels in the flow of blood in the brain. This indicates when activity in specific parts of the brain increases. In short, EEG is good for knowing when activity happens in the brain and fMRI is good for knowing where in the brain activity happens.

As a result of such research we are starting to get a deeper understanding of how the brain functions when it is making decisions. This is an area of great interest to sales professionals, because our job is to influence people and persuade them to make decisions and take action. The better able we are to understand how the brain functions when making decisions to take action, the better able we are to understand how to tailor our sales approach, messages and behaviour to achieve the results we want for our clients and ourselves.

We may think that our clients and prospective customers are intelligent, rational individuals who make well-considered and logical decisions. We may think that they go through some process of contemplating and considering the features and benefits of the product or service on offer and process this information in a logical manner in order to arrive at the decision to proceed. When our clients and prospective customers don't make the decision we want them to make we may consider them to be mistaken and foolish! We know full well what we would have done in their situation. The answer was obvious – if only they were as considered and rational as we so obviously are!

However, neuroscience research sheds new light on to how people actually make decisions, and the truth may shock you:

> According to cognitive neuroscientists, we are conscious of only about 5 per cent of our cognitive activity, so most of our decisions, actions, emotions, and behavior depends on the 95 per cent of brain activity that goes beyond our conscious awareness.
>
> (Szegedy-Maszak, 2005)

The vast majority of human thinking (including decision making) takes place below the level of conscious and controlled awareness – in our unconscious (or as it is sometimes referred to our subconscious) mind.

Let us define what is meant by the conscious and unconscious mind. Your conscious mind is your reasoning, objective level of mind. It is the mind you are aware of when you are fully awake. It is the mind that you consciously 'think' with – you are aware of your cognitive (thinking) processes. The unconscious mind consists of the processes that occur automatically and are not usually available to self-examination or meta-cognition ('cognition about cognition' or 'knowing about knowing' or 'thinking about thinking') of our thinking processes. These can include thought processes, memory and motivation. Your unconscious mind is your automatic, subjective level of mind. It operates below your level of conscious awareness. It is a mixture of thoughts, emotions, feelings, memories and other cognitive processes that we are not aware of and can probably not explain or articulate. You may at times be vaguely aware of this mental activity that exists outside your conscious awareness. A hunch or intuition that you struggle to articulate would be an example. And it is such hunches that may make the difference between customers saying yes or saying no to your sales proposal.

In this book the unconscious mind, or rather more accurately the cognitive unconscious, is defined as all of the mental processes that operate outside of conscious awareness.

To illustrate how the conscious and unconscious mind operate, our senses are receiving and taking in over 10 million bits of information every second! Our conscious brain can process only 40 bits of information per second. The rest has to be processed unconsciously. The unconscious mind will rapidly process these using an instinctive good/bad short cut interpretation that allows it to pay attention if required to anything that might threaten or assist survival and well-being.

This unconscious processing influences feelings, decision making, behaviour and actions, indeed the vast majority of thoughts and feelings that influence your customer's behaviour, and decisions about whether to purchase your product or service occur in the unconscious mind:

> At least 95 per cent of all cognition occurs below awareness in the shadows of the mind while, at most, only 5 per cent occurs in high-order consciousness.
>
> (Zaltman, 2003)

In addition, emotions are an integral part of people's decision-making process. As we will see in the next chapter, although different areas of the brain are largely responsible for processing emotions and more logical information, these areas communicate with each other and jointly influence our decision making. Emotion and reason are intertwined elements of our decision-making process. They influence and are influenced by each other.

As you will discover in the next chapter, the emotional centre of the brain is one of the oldest parts of the brain in evolutionary terms and as a result exerts the primary influence on our thinking and decision-making processes. 'Most of what we do every minute of every day is unconscious' (Paul Whelan, neuroscientist, University of Wisconsin).

So we now have a greater understanding of what is happening inside people's brains when they make buying decisions. We are beginning to understand the hurdles and challenges the brain presents to sales professionals.

Going forward we will explore how to make our sales approach and process 'brain-friendly' to ensure that the buyer's brain, at both a conscious and an unconscious level, is open and receptive to our sales message and responds positively to it. The 'Neuro-Sell' process that follows is a brain-based and brain-friendly approach to selling successfully.

But before we do that I need to take you on a guided tour inside your customer's brain – or rather your customer's three brains – which is what I'll do in the next chapter.

3
A guided tour of your customer's three brains

If the human brain were so simple that we could understand it, we would be so simple we couldn't.

(LYALL WATSON, AUTHOR)

In order to understand how the brain functions when we are selling to it, it is necessary to understand something about its structure and how it operates. Whilst this book has been designed to be highly practical in nature, your ability to successfully implement the brain-friendly selling strategies it contains will be enhanced if you have a working knowledge of the human brain. This chapter will help you to have just that.

Our brain is vital to our existence. It regulates involuntary activities such as breathing, digestion and heartbeat. It also serves as the seat of human consciousness, storing memories and enabling us to experience emotions. Our brain allows us to survive. In addition, as we will explore in Chapter 6, it gives us our personalities and makes us who we are.

The brain is very complex – indeed as mentioned in Chapter 2 it is the most complex structure in the known universe! As a result the neuroscientific language about the brain is also extremely complex.

If you are reading this book then you probably aren't a neuroscientist (hello and a warm welcome to any neuroscientists who are reading this book!) but are someone who is interested in understanding how the brain functions and operates when it makes decisions to take action so that we can use this insight to sell easily and more effectively to the brains that we interact with. So what follows is a simplification of a very complex brain. The brain is an incredibly complex and interconnected series of networks (the brain has the equivalent of 200,000 miles of 'wiring') with incredible capabilities. It contains over 100 billion brain cells called neurons, between 100,000 and 1,000,000 different chemical reactions are taking place inside it every minute, and it is capable of making approximately 200 billion calculations per second. New discoveries about it, how it works and what it is capable of are being made on a regular basis.

For practical purposes this book will contain many oversimplifications. For example, if I say that 'This part of the brain is responsible for X', then it must be remembered that no part of the brain acts alone or solely does one thing. All of the thoughts, emotions and actions we have are the result of many parts of the brain working and acting together.

Let us make sure we have an effective working knowledge of the brain's structure so that we can discover how to sell to it most effectively.

Your customers (and customers-to-be) do not have one brain – they have three brains (see Figure 3.1):

1. The old brain – this comprises the brainstem and cerebellum and is referred to as the 'reptilian brain', the 'lizard brain' or the subcortical brain. This is the oldest part of the brain (in evolutionary terms). It connects the brain with the spinal column. For ease of understanding and memory this will be referred to as the reptilian brain in this book.
2. The mid-brain – this comprises the limbic system (which is described in detail a little later in this chapter) and can be referred to as the 'mammalian brain', the 'emotional brain' and the 'truthful brain'. For ease of understanding and memory this will be referred to as the emotional brain in this book.
3. The new brain – this comprises the cortex and neocortex and can be referred to as the 'human brain' or 'rational brain'. For ease of understanding and memory this will be referred to as the rational brain in this book.

FIGURE 3.1 Your customer's three brains

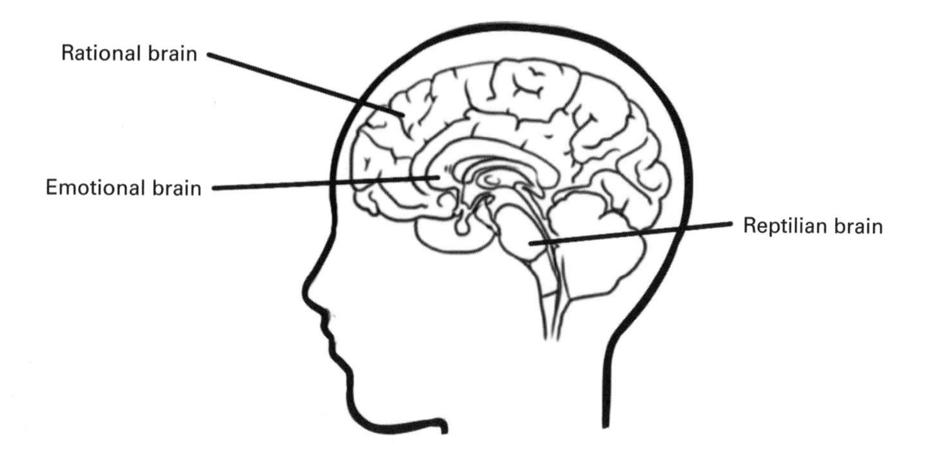

The reptilian (old) brain

The old brain gets its name from the concept that it evolved first in the evolutionary history of animals on the planet. It evolved more than 500 million years ago, and it is similar to the entire brain of reptiles today. This is the reason that it is referred to as the reptilian brain. And, although our brain evolved over time, the basic elements of it are still intact and form the lowest of the three tiers of our brain today. This primitive part of the brain is concerned with survival. Various cells in the brainstem determine the overall alertness level of the brain and regulate vital body processes such as heartbeat and breathing.

You may at this stage be wondering what this part of the brain has to do with selling. The answer is: a great deal!

Firstly, all motor and sensory nerves go through the brainstem to the rest of the body, and it plays a vital role in basic attention arousal and consciousness. For example, there is a bundle of densely packed nerve cells located in the central core of the brainstem called the reticular activating system. It runs from the top of the spinal cord into the middle of the brain. In addition to a host of other functions the reticular activating system is the brain's chief gatekeeper, which screens and filters what type of information will be allowed through. Anything that is deemed as irrelevant is filtered out. It is rather like a PA or secretary who decides which calls get put through to the CEO.

Only two categories of information are allowed through: 1) information valuable to have right now; 2) information that alerts you to threat or danger. This primitive part of the brain has little or no patience if the subject does not immediately concern well-being and survival. At first contact with a stranger, for example (such as a first meeting with a new sales professional such as yourself), it will instantly conduct a threat response and decide if the stranger is friend or foe. It will also determine if the stranger could be a form of sustenance or a potential candidate to reproduce with!

It prioritizes survival first (the avoidance of pain and danger) and then achieving comfort (so it will respond to pain avoidance first). It should be stressed that this is a mechanical, selfish and unconscious part of our brain.

It does however have a very strong influence. For example, if initial contact with a salesperson stresses the 'gatekeeper' the automatic fight/flight/freeze response is stimulated. This can happen in a fraction of a second. Part of this process includes shutting out all other message receptors, which means your opportunity to communicate is severely limited.

Several of the characteristics of this part of the brain will be referred to in later chapters as we look at how to make this part of the customer's brain regard us as friend (rather than foe), and to classify ourselves and what we are selling as useful and rewarding and therefore something to be paid attention to.

The emotional (mid-)brain

As the brain evolved the mid-brain or limbic system developed. It is referred to as the mammalian brain, as it is thought to have first evolved in mammals. This is where emotions are generated, along with many of the urges (usually concerned with survival) that direct our behaviour. The limbic system has other functions also. For example, a part of the brain called the thalamus acts as a relay station directing incoming sensory information to the appropriate parts of the brain for further processing.

It is important to realize that although this part of the brain is also unconscious in function it has a profound effect on us because it links the brainstem with the higher reasoning functions of the cerebral cortex, and feeds information to it.

The limbic system is a part of the brain that in a similar way to the reptilian brain reacts reflexively, instantaneously and without thought in real time.

It gives off a true response to information coming in from the environment and plays a key role in developing and carrying out instinctive emotions and accompanying behaviours. For that reason it is sometimes referred to as the 'truthful brain'.

In terms of behaviour it is also the part of the brain that generates our body language, and as we will see in Chapter 14 a sales professional's ability to read and respond according to the customer's body language is a powerful skill to master and an important part of the 'Neuro-Sell' brain-friendly selling process.

A very active element of the limbic brain is what we can refer to as the 'fear system'. This system detects danger and instinctively produces reactions and behaviour that will maximize your chances of survival. The key part of the brain that is involved is called the amygdala. These are small regions (your brain possesses two amygdale – one in each hemisphere) in the forebrain where fear is registered and generated.

Information about external stimuli reach the amygdala via a direct pathway from the thalamus (the brain's relay station mentioned earlier), as well as travelling via the part of the brain called the cortex, which will be described shortly. As you may expect, the direct thalamus-to-amygdala route is faster than if the information goes via the cortex first. In survival terms this is advantageous, as it allows us to begin to respond to the perceived danger before we know fully what the stimulus is.

If the information travels directly to the amygdala it misses out on the benefit of cortical processing and will at best be a crude representation of the stimulus. Most of us will have had the experience of seeing something that our limbic system perceived as a threat (there is a snake in our garden shed!), which triggers a fear reaction (for example, our heart rate suddenly increases) and then once the stimulus has been assessed turns out not to be a threat – the snake turns out on closer examination to be a piece of rope! And relax...

Again you may at this stage be wondering what this part of the brain has to do with selling. The answer (again) is: a great deal!

> The amygdala has a greater influence on the cortex (where rational, analytical thinking takes place) than the cortex has on the amygdale, allowing emotional arousal to dominate and control thinking.
>
> (Professor Joseph LeDoux, neuroscientist)

The phrase 'emotional arousal... dominate[s] and control[s] thinking' is of great importance to us as sales professionals. The limbic system can dominate and control the thinking of your customer.

As mentioned in Chapter 2, the majority of cognition including decision making is unconscious. To sell effectively we must make sure that we and our sales messages are 'brain-friendly' so that we can arouse the limbic emotional brain in a way that supports our selling rather than handicaps it. This concept is at the very heart of the 'Neuro-Sell' brain-friendly selling process.

The rational (new) brain

The cortex and neocortex are the newest (in evolutionary terms) parts of the brain. Because it is responsible for complex thought this part of the brain is sometimes referred to as the 'thinking brain' or the 'intellectual brain'.

It is this part of the brain with its ability to analyse and interpret data at a level that is unique to human beings that sets us apart from the rest of the animal kingdom. This part of the brain processes information received from the senses and regulates cognitive functions such as thinking, speaking, learning, remembering and making decisions.

Although the cerebral cortex of the brain is complicated it basically fulfils four key functions:

1. *Sensing.* This is the receipt of sensory signals from the outside world. Each of the five senses picks up signals and sends them to specific regions of the brain for each sense. The signals come into the brain as individual pulses of electrical energy from each of the sensory organs, and these small bits of information have no meaning to the brain in their raw form. In short, it is getting information.
2. *Integration.* This is where the individual signals get added together. The small bits are merged into larger patterns that become meaningful such as language and images. In short, it is making meaning of this information.
3. *Creating ideas and plans.* When the parts have been integrated the sum of them generates a plan for what action is required and where. In short, it is creating new ideas.
4. *Execution.* The motor function then executes these plans of action by sending motor signals to the muscles, which act in coordinated ways to create the required movements. In short, it is acting on the ideas.

If you were to look at the brain from above you would see that it has two hemispheres (which are divided by the longitudinal fissure) (see Figure 3.2), and it is covered in a thin skin of folded and wrinkled tissue called the cerebral cortex. If unfolded the cortex would measure about 60 centimetres by 60 centimetres. The two cerebral hemispheres account for approximately 85 per cent of the brain's weight.

FIGURE 3.2 The two hemispheres of your customer's brain

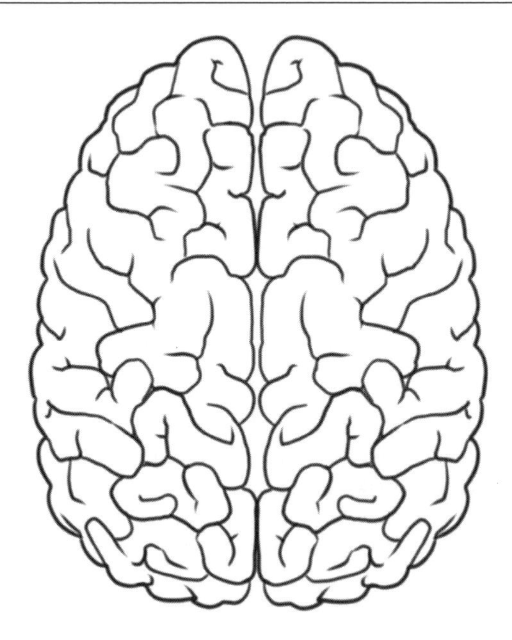

The billions of neurons in these two hemispheres are connected by bundles of nerve cell fibres called the corpus callosum, which constantly transfers information backwards and forwards between them. Information arriving in one hemisphere is almost instantly available to the other hemisphere, and their responses are closely harmonized to provide what appears to be a seamless perception of the outside world. Some research suggests that the corpus callosum is more developed in the female brain, with many more neurons connecting the left and right hemispheres, enabling the female brain to move between the processing capabilities of each of the hemispheres more easily than the male brain.

Although not identical, to a large extent each of the hemispheres is a physical mirror image of the other. The function of the hemispheres has been the subject of much interest and research by neuroscientists. Each hemisphere appears to possess its own special skills, strengths and weaknesses, and its own way of processing information.

Perhaps as a result of the excitement and interest generated by research into the difference between the left and right hemispheres, an oversimplified view of hemispheric functioning emerged and became popular. A series of far-too-rigid dichotomies emerged that ascribed various functions solely to the left (logic, science, reason, cognition) and right (art, creativity, intuition, emotion) hemispheres. While neuroscience has moved far beyond this, the oversimplification still persists in the wider public perception. The concept of such a rigid divide between the functions of the left and right hemisphere is a myth, and a popular myth at that!

Brain regions have certain functions because of the integrated brain systems that they are a part of. These functions are properties of these integrated systems rather than of isolated areas of the brain. Mental functions involve interconnected regions of the brain acting together.

That being said, research shows that the two hemispheres do have specific 'hard-wired' functions that under normal circumstances will always tend to develop in a particular hemisphere. The brain is very malleable or plastic and can be influenced by environmental factors, so some brains do end up being organized in very different ways.

Although the concept of such a rigid divide between the functions of the two hemispheres is too simplistic, and the constant interaction between the hemispheres makes it challenging to pinpoint what is happening where, brain imaging shows us that the suitability of the hemispheres to specific functions or skills is approximately as is popularly understood – although these functions and skills are not as rigidly divided to one hemisphere or the other as many people believe. Under normal circumstances certain skills will tend to develop on a particular side of the brain. It is believed that the different processing styles and structure (there are subtle differences between some neurons on the left and right side of the brain) of the two hemispheres influence the various functions that they tend to perform.

The left hemisphere is often described as analytical, as it is suited to recognizing the parts that make up the whole. It appears to specialize in linear processing, which is sequential, moving from one point to another in a step-by-step manner. It is analytical, logical and precise. This makes it suitable for thinking about and implementing detailed plans. It is seen to be more calculating than the right brain and has an affinity for whatever is mechanical and impersonal. It needs certainty and to be right.

The right hemisphere is often described as holistic, as it is suited to combining the parts to make a whole. It appears to specialize in parallel or simultaneous processing, integrating individual parts or components and organizing them into a whole. It seeks patterns and gestalts (an organized whole) and is interested in relationships. It is seen to be more emotional than the left brain. It has an affinity for whatever is living and personal. The right hemisphere makes it possible to hold several possibilities simultaneously, and it is able to tolerate uncertainty. Abilities such as empathy and self-awareness are largely dependent upon the right hemisphere. In general it is more closely connected with the limbic system and is involved in the experience of emotion.

> The right hemisphere is predominantly hard-wired for empathy. The left hemisphere is predominantly hard-wired for systemizing.
>
> (Professor Simon Baron-Cohen, Professor of Developmental Psychopathology, University of Cambridge)

> The right hemisphere is particularly adept at processing novel information and the left hemisphere is particularly adept at processing routine, familiar information.
>
> (Professor Elkhonon Goldberg, Clinical Professor of Neurology, New York University School of Medicine)

Each of the hemispheres is divided into four cortical lobes (occipital, parietal, temporal and frontal). The frontal lobes are divided from the parietal, occipital and temporal lobes by the central sulci and the lateral fissure. The regions of the brain that predominantly receive sensory information are located behind the central sulcus and lateral fissure in the occipital, parietal and temporal lobes. The occipital lobes lie at the back of the brain and are made up almost entirely of visual processing areas. The temporal lobes are situated around the ears and deal with sound, speech comprehension (usually in the left hemisphere only) and some aspects of memory. The parietal lobes sit above the occipital lobes and deal with functions connected with movement, orientation, calculation and certain types of recognition. In front of the parietal lobe sit the frontal lobes, which deal with thinking, conceptualizing and planning.

There is a functional difference between the front and back of the cortex. Sensory information from the outside world goes predominantly to the sensory cortex in the back of the brain. This part of the cortex is involved in long-term memory and where the brain maps our knowledge of the world. It contains data from the past and is where connections between different experiences are made.

This large region at the back of the brain (encompassing the occipital, parietal and temporal lobes) is not just an area for processing sensory information; it is also where information from the various senses is associated and integrated together for higher-order processing.

The rear half of the brain is described by *PRISM* Brain Mapping (which will feature extensively in Chapter 6) as the 'database brain'.

The front of the cortex is where some of the most advanced functions of the brain are performed. These are sometimes referred to as the 'executive functions' of the brain. It is here that thoughts are organized so that they make sense, things are weighed and considered, decisions to take action (or not) are made, plans are developed and progress is monitored. This is the part of the cortex that is active in creating ideas and solving problems. The front of the cortex is more orientated to the future.

A key section of the brain involved in the above activities is called the prefrontal cortex. This is a section of the outer layer of the brain that sits behind the forehead. It was the last major region of the brain to evolve. Although it accounts for approximately only 5 per cent of the brain's volume, it is the key part of the brain that gives human beings such an advantage as a species. The prefrontal cortex coexists with the limbic system in a delicate balance. It is the prefrontal cortex that acts to restore the balance when emotions get stirred up and potentially out of control. In times of emergency and great stress the emotional limbic system takes control of the brain.

Although we live in modern times, our brain has not changed significantly for 100,000 years and at an unconscious level can treat modern-day stressful situations that are clearly not life-threatening as a threat. When this happens the limbic system becomes dominant and the rational cortex is not able to function as effectively as would be helpful.

As we shall see as we go forward this can be very important to our success as sales professionals in closing the sale!

Mirror neurons

Now that you have an understanding of your customer's three brains, I would like to explore one of the most fascinating discoveries in neuroscience – mirror neurons. They were discovered by neuroscientist Dr Giacomo Rizzolati from the University of Parma in Italy. Dr Rizzolati and his team were conducting experiments in motor neurons (motor neurones are neurones that carry signals from the spinal cord to the muscles to produce movement) and were being helped by some monkeys whose brain activity was being monitored. One day a lab assistant returned from a break eating an ice cream. A monkey who was also taking a break from participating in experiments was just sitting in a relaxed manner. As the monkey observed the ice cream being consumed, electrical activity in its brain was triggered as though it was actually consuming the ice cream. This included making the physical movements required, for example lifting its arms to raise the ice cream to its mouth despite the fact that the monkey was not eating but only watching.

Rizzolati's interest was aroused, and his team developed a series of studies. In one study, when a monkey saw another monkey or a human eat a peanut, the neurons in the monkey's brain fired as if it was also eating the peanut. Time after time, neurons in the prefrontal cortex reacted to the *perception* of the actions that were observed.

The theory about mirror neurons is that, when you watch someone perform an action, for example participating in a particular sport, you automatically simulate the action in your brain. However, it is not just the physical action that is simulated. Mirror neurons appear to be able to reproduce or mimic almost anything we experience, including the emotions another person is feeling. If you observe a group of sports fans watching their favourite team you will see them physically tensing up, moving, wincing, cheering and smiling as their mirror neurons respond to the action on the pitch. They respond almost as if they were playing the sport themselves.

To continue the sporting theme, interestingly our mirror neurons react when we see a ball being kicked, when we just hear a ball being kicked and even when we say the word 'kick' or hear it being said.

When watching a movie in the cinema or a play in the theatre we see actors use emotions to express how their characters are feeling, and our mirror neurons inspire the same feelings in us, moving us emotionally to be excited, thrilled, sad or happy.

On the PBS television programme *NOVA scienceNOW* (PBS, 2005), presenter Robert Krulwich participated in an experiment with Professor Marco Iacaboni from UCLA where he looked at a series of photographs of different facial expressions whilst having his brain scanned by an fMRI machine. In the first part of the experiment Krulwich was asked to physically mimic the facial expressions he saw. In the second part of the experiment Krulwich was asked just to look at the photographs of the facial expressions and remain motionless.

The results showed that the part of Krulwich's brain that was activated when he made a facial expression was also activated when he only saw the facial expression but didn't mimic it. In addition when he saw a happy face the 'happy emotional part' of his brain activated, even when he made no facial expression himself. When he did mimic the facial expression the relevant part of the brain became even more active. Professor Iacoboni believes that mirror neurons can send messages to the limbic system and enable us to tune into, empathize with and connect with each other's feelings.

Mirror neurons are also believed to be a very powerful learning system, where we can rapidly learn from others as mirror neurons respond when observing them performing certain behaviours. Mirror neurons are sometimes referred to as the 'monkey see monkey do neurons'.

A further theory is that mirror neurons are a powerful predictive survival system, as articulated by Dr Giacomo Rizzolati: 'Our survival depends on understanding the actions, intentions and emotions of others... Mirror neurons allow us to grasp the minds of others not through conceptual understanding but through direct simulation. By feeling, not by thinking' (Blakeslee, 2006).

So we have seen that the human brain possesses multiple mirror neuron systems that specialize in understanding people's actions, the social meaning of their behaviour, their emotions and their intentions.

Professor Iacaboni strongly believes that mirror neurons provide a unifying mechanism that allows people to connect at a simple level. As he states, 'Mirror neurons suggest that we pretend to be in another person's mental shoes. In fact, with mirror neurons we do not have to pretend; we practically are in another person's mind' (Than, 2005).

We will return to the fascinating subject of mirror neurons and how they may apply to selling effectively in Chapter 9.

This concludes our tour of your customer's three brains, and many of the themes from this chapter will be referred to as this book progresses. In the next chapter we will explore the process our customers (and their brains) go through when they buy something.

4

The buying process and the buying brain

There is a process involved in selling successfully. Indeed selling is a process. Some of the clients that I work with have a structured sales process that their sales leadership encourage and train their salespeople to use.

A more common situation I encounter is that the clients I work with don't have a sales process (or at least they don't have until I get involved!). Their salespeople do the best that they can and to some degree they 'make it up as they go along'. They will have some approach that they follow, but it is likely to be largely unconscious and based upon historical trial and error. Although there will be successful salespeople within companies that have not adopted a structured sales process, their success will be more as a result of accident than design. In general, an underperforming sales force is the result of all categories of salespeople from underperformers to those at the top of the sales league table never fulfilling their true sales potential.

The degree of success that salespeople experience is often directly related to their ability to follow a tried and tested and proven sales process. When a proven sales process is followed correctly the result is increased sales.

In Chapter 8 you will be introduced to a cutting-edge 'brain-friendly' selling process – but more of that later!

An even more concerning situation is that I can count on the fingers of one hand the number of clients I initially encounter whose sales process is orientated around and to the *customer's buying process*. If we pause for a moment to consider then this can be seen to be a concerning situation. The salespeople will be orientating their sales process (if indeed they have a conscious process) to their own aims and agenda. They will be largely viewing the sales process from their perspective. It is a sales process that they, in some way, take the customer through. The focus is largely on the result that the salesperson wants to achieve.

This is concerning because the customer is the most important person in the sales interaction. For a start, customers are the ones with the money! It is customers who will make the final decision whether to buy from you or not, and they will do so only if they believe that the purchase will benefit them in achieving their aims and objectives. It would therefore make sense to consider things from their perspective, wouldn't it?

I am aware that there will be some experienced sales professionals who are reading this and getting somewhat offended that I am implying they are not customer focused. I am not implying that you are not customer focused – I am implying that you are not customer focused enough!

If you don't do so already, I am going to invite you to consider the sales process in terms of the customer's buying process. Let us assume the customer's perspective, identify the process the customer will be going through when moving towards a purchase decision and then *align our selling process to match the customer's buying process*. In doing this we will be providing customers with whatever it is they need to move through *their buying process* to a successful conclusion. When this is done well the successful conclusion will usually involve making the decision to purchase from you.

So, if successful sales professionals orientate their sales process so that it aligns with and follows the customer's buying process, then what does the customer's buying process look like? To illustrate let me use an example that most people will have experienced of purchasing a new electrical item such as a television, laptop, tablet or smart phone. This is a reasonably significant purchase for most people and therefore will usually not be a rapid, impulse purchase (although there are always exceptions!).

Firstly, you identify that you have a problem. For example, your current television breaks down, your laptop's performance starts to slow down or you become aware that your current phone is now looking rather dated as against the newer models your friends possess. You will then go through a process of identifying a possible solution to your problem. You may conduct some research online, read some magazines that contain useful information, ask your friends for their advice, visit one or more retailers to browse or ask the salespeople in a store for their advice. Depending on the purchase and your personality (more on this later!) this may be a short or a long step in your buying process. Using the information you have gleaned, you will then refine this and identify your preferred solution to your problem. For example, the model of smart phone you want is decided or the specification of your new tablet is defined. You will then identify potential suppliers and may request pricing quotations from them and compare and contrast these to determine where to place your business. You may choose to use the information you have gleaned to negotiate with more than one supplier, perhaps playing one off against the other. You will then make your final buying decision and purchase your desired item. Once you have taken possession you will evaluate your purchase against the criteria or specification you decided you wanted.

This is a fairly simple and straightforward process, which most of us will have been through as consumers many times. You may also have experienced, as I have a number of times with well-known high street electrical retailers, being at a relatively early stage in the buying process (such as gathering information on possible options), only to encounter a salesperson in the store whose sole focus seemed to be to get you to buy something as fast as possible – perhaps by pushing you towards that week's 'never to be repeated special offer'. The salesperson was clearly much further ahead in your buying process than you were, which is very annoying and, although driven by an understandably strong desire to get the sale closed as quickly as possible, very unprofessional.

At an individual level it will vary from person to person how conscious and considered the buying process is. For example, you will recall from Chapter 2 that up to 95 per cent of brain activity, including decision making, takes place beyond conscious awareness. However, I would suggest that the process outlined will, to some degree, be followed for significant purchases.

Where larger-scale buying (in terms of the amount of money involved or the strategic importance of the purchasing decision) is taking place then a more delineated and formal procurement process will usually be followed. The common steps are not dissimilar to those of the personal buying process outlined above and will usually consist of some variation of the following:

1. Identify the need or problem or become aware that a need or problem exists.
2. Identify possible solution characteristics.
3. Search for possible suppliers of a solution.
4. Request solution proposals from one or more suppliers.
5. Analyse the solution proposals.
6. Evaluate the solution proposals to determine the preferred supplier.
7. Negotiate terms with the chosen supplier.
8. Award the contract.
9. Integrate and induct the supplier.
10. Review the supplier's performance.

If customers have an identified and structured buying process then they will usually share this with you. Your task then is to align your sales process to customers' buying processes, providing them with exactly what they need at the time they need, therefore making the sale much easier. An awareness of customers' buying processes allows you to get ahead of the game and be well prepared as each stage presents itself.

If, as is often the case, customers do not have a formal or conscious buying process then it is important to elicit the process they will go through. As 95 per cent of brain activity is unconscious, customers may have an unconscious or instinctive process they will follow, and it can be helpful to both them and you if you can make this more conscious. Asking questions to elicit their (largely) unconscious process will be useful. Examples include:

- 'What process will you follow to make your decision?'
- 'How will you be making your decision about what is right for you?'
- 'How will you go about selecting the right solution or supplier?'

Helping customers to get clarity about how they will make their purchasing decision will add value to the customers and help to position you as a credible, helpful and useful supplier.

Go upstream!

It is best practice to become involved in the buying process as early as possible. If, for example, you can get involved in the first stages of need or problem identification and identification of possible solution characteristics then you will be best placed to influence and shape customers' understanding of what they need. We will return to this subject when we look at the 'Neuro-Sell' brain-friendly selling process in Chapter 8.

Having defined a typical buying process we can now go one step further and integrate what we understand about how the brain works when it is making buying decisions.

A primitive brain in a modern world

To fully understand how the brain operates when making buying decisions it is important to realize that approximately 100,000 years ago our brain reached its current form and size. Our brains, though highly adaptable, haven't developed or changed a great deal in the last 100,000 years, so although we live in a modern (compared to our primitive ancestors) and very safe world a lot of our brain function (particularly our unconscious brain functioning) is designed to ensure our survival, as was outlined in Chapter 3.

Our brain is attuned to achieving this task. It evolved at a time when food could be scarce, and if you wanted to stay alive you had to be able to hunt effectively, rapidly spot danger, and distinguish between friend and foe. In addition, you wanted to ensure the ongoing survival of your species, so finding a suitable mate and reproduction were also high on the agenda – so in some ways not a lot *has* changed!

This has great ramifications when we are considering how best to sell to our customer's brain. As mentioned earlier, most of our decisions, actions, emotions and behaviour depend on the 95 per cent of brain activity that takes place beyond our conscious awareness. Although we may like to consider ourselves (and our customers) as intelligent, rational and logical thinkers and decision makers, in reality most of the thoughts and feelings that exert an influence on us occur in the more primitive, unconscious areas of our brains. Whilst we (and our customers) may like to think that we make very considered and conscious decisions, and that we follow a structured process where we carefully consider the key attributes and criteria a product or service may deliver to us, this forms only a relatively small part of how we make decisions.

We need to be aware that in the human brain decision making occurs at two (at least) levels – in the more evolved and rational cortex and in the more primitive subcortical area in the limbic and reptilian brain.

> In reality, people's emotions are closely interwoven with reasoning processes. Although our brains have separate structures for processing emotions and logical reasoning, the two systems communicate with each other and jointly affect our behaviour. Even more important, the emotional system – the older of the two in terms of evolution – typically exerts the first force on our thinking and behaviour. More important still, emotions contribute to, and are essential for, sound decision making... Indeed, decision making hinges on the simultaneous functioning of reason and emotion.
>
> (Zaltman, 2003)

The older parts of the brain exert a powerful influence on our behaviour. Our brains process the vast majority of the data received from our senses unconsciously, and the primitive parts of the brain can and do react unconsciously to stimuli in a matter of milliseconds. This is much faster than the rational cortex. At its core the brain is very emotional, the cortex and the limbic system coexist in a delicate balance and the immense power of the more primitive parts of the brain cannot be overlooked if we want to sell successfully to it. The limbic system will react faster and before the cortex and therefore can hugely influence how the brain is functioning and therefore how receptive it will be to your sales message.

In Chapter 8 we will discover how to approach a customer in a 'brain-friendly' manner so that the brain accepts, rather than rejects, us and our sales message.

Stay away from danger; move towards reward

A fundamental organizing and operating principle of your brain that drives your thinking, behaviour and action is to avoid and move away from anything that is perceived to be painful, dangerous or threatening, and to move towards anything that is pleasurable, comforting or rewarding. At its core this is a hard-wired survival instinct of the human brain. It has played a vital role in our evolution, and although it may not always be as practically useful in our safe, modern world it is still driving our behaviour and that of our customers.

> Everything you do in life is based on your brain's determination to minimise danger or maximise reward. Minimise danger, maximise reward is the organising principle of the brain.
>
> (Gordon, 2000)

Although the 'stay away' and the 'towards reward' drives are active all the time in the human brain, it is the 'away from' drive that is stronger and faster. If we consider the almost constant danger our primitive ancestors experienced in their environment it makes evolutionary sense to prioritize keeping us safe from injury or loss of life by giving this drive a stronger limbic response. It is very rare for the vast majority of people in our modern society to be exposed to genuine threat of injury or death. However, the hard-wired 'away from' survival instinct is still constantly active.

Situations or stimuli that appear to be or are coded by the brain as threatening can also trigger the 'away from' response. These include a loss of control, loss of approval, loss of status or standing, walking into a room of strangers, speaking in public, change that is imposed upon us, losing our jobs and, in the case of a customer when making a buying decision, fearing that a salesperson may con or trick us in some way into making an unwise buying decision. In addition, meeting a stranger for the first time can trigger a threat response in the human brain.

Although few of these scenarios are truly threatening in terms of injury or death the reptilian and emotional parts of the brain do not have the capacity for logic and analysis. It is instinctive and automatic, and many stimuli will be categorized as posing a threat by this part of the brain. Speaking in front of a large group of people is an example that many people will be able to identify with.

If the threat response is triggered, resources of glucose and oxygen will be drawn away from the customer's prefrontal cortex, making it more difficult for the customer to make decisions and think about new ideas and concepts. The customer's brain is more likely to revert to automatic or unconscious processes and err on the side of caution in order to minimize risk.

Rewards that the brain is interested in will include food, sex, relationships and connection with others, approval from others, satisfaction from achievement (the brain is goal orientated), certainty, happiness and contentment. On a moment-by-moment basis your brain is constantly scanning your environment, looking for signs of danger or opportunities for reward. So as your customers proceed through the various stages of the buying process their brains will be constantly seeking signs of perceived danger or reward.

At a conscious and particularly unconscious level customers' brains will be moving away from pain and towards reward questions such as: 'Does this ease my pain, solve my problems, ease my frustrations, reduce my stress, keep my job safe, get my boss off my back, stop me getting sacked or stop me looking stupid, incompetent or incapable?' 'Does this bring me pleasure, make me look good, get me approval, get me some more time, make me more money, help me to achieve results, help me make a wise decision, make me positively recognized, help me achieve my targets, get me my bonus or get me promoted?' We will return to using these two powerful driving forces to close more sales in Chapter 11.

Neural maps

As described earlier, our brains contain somewhere in the region of 100 billion nerve cells called neurons (see Figure 4.1), which create brain activity by carrying an electrical signal from one neuron to another. Each neuron can connect with up to 10,000 neighbouring neurons. The parts of the neurons that join up are rather like branches, and there are two kinds – axons, which conduct signals away from the neuron, and dendrites, which receive incoming information. The part of the neuron that does the connecting is called the synapse – each neuron can have many synapses. Information is communicated across a tiny gap where each axon meets a dendrite, called the synapse.

FIGURE 4.1 A neuron

In order for electrical signals to cross the synaptic gap, each axon secretes brain chemicals called neurotransmitters, which are released into the gap when the neuron is suitably 'fired up'. The neurotransmitters trigger the neighbouring neurons to fire up too. A chain-like effect is produced simultaneously in millions of neurons that are connected together.

The brain is an incredibly busy place, with 100 billion neurons and 100 trillion synapses communicating information with each other. The synapses that communicate with each other the most frequently, communicating the same message repeatedly, become stronger. The synapses that communicate less frequently become weaker. The connections that are made between neurons can be thought of as pathways. Information flows quickly and easily along old-established pathways. A fresh pathway information flow will be slower and more difficult and use more energy. This is why learning a new skill takes more effort than performing a skill you have mastered. If you are an experienced driver you will find driving your car to be a relatively effortless task. If you are a learner driver you may find yourself exhausted at the end of a one-hour driving lesson. The brain is designed to be energy-efficient, so established pathways and automatic processes will tend to become the brain's default, as they use less resource than establishing new pathways.

Despite the commonly used metaphor the brain is nothing like a computer! It does not process information or assemble thoughts and feelings from individual bits of data such as strings of digits. The myriad of neural pathways in the brain combine to create mental frameworks or patterns that we can usefully think of as 'maps' – they are sometimes referred to as 'schemas'. These 'maps' or arrangements of neurons represent entire 'chunks', concepts or templates of knowing, perceiving or cognition. Most cognitive functions involve the interaction of such maps from many different parts of the brain at once. The brain assembles perceptions by the simultaneous interaction of whole 'maps', relating one whole map to another and looking for similarities, differences or relationships between them.

At one level these can be thought of as mental structures of preconceived ideas, organized patterns of thought or a framework that represents various aspects of the world we inhabit. They are very useful, as they allow us to navigate through our everyday environment without effortful processing – this uses minimal brain resources, saving energy for other tasks. For example, everybody has a map of how doors work. The map will be a generalization about the function of doors (to facilitate our entry to and exit from buildings, for example) and how they operate (they move in one of two directions, inwards and outwards, and entry into the desired building is by either pushing or pulling as appropriate). The appropriate action can be determined by either observing other people using the door in advance of you or by looking for helpful signs that indicate the appropriate action required. This very useful map allows us to move in and out of the buildings we inhabit using the minimum amount of brain resource. Imagine how much mental energy you would use if you consciously had to assess and analyse every door that you encountered on a day-to-day basis!

We use our maps to organize our current knowledge, which provides us with a framework for future understanding. The maps provide us with a system of organizing and perceiving new information. The way we understand something new (a new door in a new building we have never visited before) is to link it into an existing map or framework. We will tend to rapidly organize new perceptions into existing maps, allowing us to act with minimal effort.

Now for the bad news about maps! Firstly, our brains are far more likely to notice things that fit into our current maps. This can influence and inhibit the uptake of new information. This proactive interference can be seen, for example, in the many generalizations and stereotypes that human beings use in the course of their day-to-day life. Some of these (as in the case of the door example) are highly useful, whereas others can dramatically inhibit people considering new ideas, ways of working, innovation and so on. The brain's default is to stick with what is known, as this consumes fewer energy resources and gives the brain a sense of certainty, which is important to a brain that has evolved to survive in a hostile and challenging primitive environment!

In addition, as mentioned above, when we are trying to understand new ideas, thoughts or concepts we will attempt to connect the new thing with our existing structure. We attempt to understand the new with reference to the content of our existing maps. This content will be gleaned from and referenced to past experiences.

So, when our customers are making a buying decision, they will be processing the potential new purchase in relation to data that are currently stored within the content of their existing map frameworks. This can make the selling of new concepts or ideas challenging. It can be difficult enough at a conscious level, but if we then add the challenge that 95 per cent of cognition, including decision making, occurs at an unconscious level the scale of the challenge is clear.

To succeed we will need to be able to 'speak' to and work with all three of a customer's brains at each stage of the buying process. The salespeople who are likely to be the most effective will work with the cortex, where more rational thought takes place, and be sensitive to the huge influence the subcortical area (the reptilian and limbic brain) has on the customer's decision-making process.

As we move through the 'Neuro-Sell' brain-friendly selling process this book will explore and expose you to how to do exactly that.

5
Adaptive selling

In the past, salespeople were often trained to follow a very standard one-size-fits-all sales presentation or sales script. The idea was that the standard presentation or script contained tried and tested selling techniques that would persuade the prospective customer to say 'yes'. This approach was symptomatic of the more traditional, transactional 'push' style of selling that was prevalent in the past. It is sometimes referred to as 'spray and pray' or 'show up and throw up'. That is, you deliver your standard sales message to every prospective customer and hope that sometimes it will get you a positive result.

As the world of selling evolved, perhaps in response to customers becoming more educated and resistant to the standard and all-too-common 'push' approach, a shift towards a more consultative and tailored approach to sales became more common. The salesperson spent more time understanding the customer's context, circumstances and challenges so that a more customized solution could be devised that would more accurately reflect the customer's unique needs. This is sometimes referred to as a 'pull' approach, as the information the salesperson requires and, to varying degrees, the structure of the solution are 'pulled' or elicited from the customer.

More and more salespeople began to be trained in consultative selling techniques, and on the whole these have proved to be more effective than the traditional 'push' approach. In addition, more and more customers have experienced the consultative approach and have come to expect it and prefer it.

Criticism has sometimes been levelled at the more consultative approach that it is a lengthier and more time-consuming approach. Salespeople using a consultative approach have been criticized for being too customer-orientated and lacking the ability to challenge their customers and drive the sale to a conclusion.

So it would appear that broadly speaking we have two approaches to selling. In my opinion debating the validity of the two approaches is short-sighted. What we appear to have is a continuum of selling styles and approaches ranging from, at one extreme, a 'push' or 'hardball' approach to, at the other extreme, a highly consultative 'pull' approach. To debate the superiority of these two extremes adopting an either/or approach is to my mind a very narrow and unproductive exercise. What will be more productive and useful is to explore an approach to selling that is practical, flexible in approach (avoiding either a one-size-fits-all or an either/or approach and indeed incorporating the best of each approach as required) and most importantly proven to improve sales performance. Allow me to introduce you to – adaptive selling!

When using adaptive selling, salespeople flex, alter and vary their selling approach depending upon:

1. the nature of the selling situation;
2. the stage of the buying process that the customer is currently in;
3. the specific interests and needs that the customer has in relation to the product or service in question;
4. the personality and buying style of the customer, as adaptive salespeople will tailor their questioning, probing, sales presentation and closing methodology based on the customer's behavioural preference, and will also respond to feedback (both verbal and non-verbal) that they receive from the individual and adapt accordingly.

Adaptive selling is a practical and powerful approach to selling. Indeed research has demonstrated that the practice of adaptive selling is welcomed by salespeople and has been shown to increase their sales performance. A research paper by James Maxham at Louisiana State University concluded that:

> In the midst of increased competition and rising training costs, management should consider incorporating adaptive sales training into their training structure. Salespeople in this study buy into adaptive selling as an effective method. It has been shown to increase sales performance, and salespeople have indicated in this study that more adaptive sales training is necessary, in relation to other training topics.
>
> (Maxham, 1997)

The theory behind adaptive selling is that the successful outcome of a selling situation is determined by both the customer's deliberation of the benefits of the product or service being offered *and* the customer's experience during the sales interaction. The salesperson's ability to create the right chemistry, rapport and connection with the customer will be as important as the ability to communicate the key benefits of the proposal to the customer.

Neuroscience research into how the brain functions would support this thinking. As we saw in earlier chapters, reason and emotion are intertwined elements of our decision-making process. They influence and are influenced by each other. Adaptive selling is an approach that acknowledges the importance of both reason and emotion with regard to selling successfully.

This chapter outlines the core elements of adaptive selling, so that you can successfully incorporate these into your sales approach, which in turn will lead to improved sales results.

Adapting to the nature of the selling situation

There are certain ways in which salespeople may have to adapt, dependent upon the nature of the selling situation they face. These may include the factors detailed in the following sections.

Industry or sector

The structure of specific industries can lead to salespeople needing to adapt their approach in order to be successful; for example, there may be specific constraints, such as legal requirements or distribution channels. In addition, specific industries or sectors will possess different characteristics, such as their relative speed or competitiveness, which will drive the need to adapt accordingly.

The size of the sale in question will also have an influence over the salesperson's behaviour. For example, a large transaction may require a more in-depth and detailed sales approach, perhaps within the formality of a structured tender process, whereas a smaller transaction may be able to be dealt with more quickly. In some industries the sales cycle will be short and in some it will be long, and salespeople have to adapt their approach accordingly.

It is not within the scope of this book to go into the intricacies of a myriad of different industries and sectors. Readers who are experienced in their industry or sector will already understand where and how they need to adapt accordingly. If you are new to your industry or sector then building your knowledge of how it operates and its structure and methodologies should be a high priority for you so that you can adapt accordingly.

If you are selling directly to the consumer rather than business to business then again a certain degree of adaptation will be required, dependent upon the nature of the product or service you are selling. Generally speaking, the higher the value of the transaction, the more care and deliberation customers are likely to take when making their purchasing decision. For 'big-ticket' purchases more consideration will have to be given to eliciting and then aligning to the buying process that the customer will follow.

I will assume that you possess the necessary knowledge (or are currently engaged in a rapid learning process if you do not!) to adapt your sales approach to your industry or sector, so that we can look at the other three key areas of sales adaptation.

The stage of the buying process that the customer is currently in

In Chapter 4 I outlined the typical stages in a customer's buying process:

1. Identify the need or problem or become aware that a need or problem exists.
2. Identify possible solution characteristics.
3. Search for possible suppliers of a solution.
4. Request solution proposals from one or more suppliers.
5. Analyse the solution proposals.
6. Evaluate the solution proposals to determine the preferred supplier.
7. Negotiate terms with the chosen supplier.
8. Award the contract.
9. Integrate and induct the supplier.
10. Review the supplier's performance.

And as previously mentioned it is very important to adapt our selling approach at each stage of the customer's buying process so that we align our selling process to the customer's buying process. If you understand customers' buying processes then you will be able to provide them with whatever they need (eg information, specification, pricing) at the stage that they need it. This will allow customers to move through their buying process as quickly as possible, which means that the deal is closed as quickly as possible. This is good news for both customer and salesperson.

In addition, if you understand customers' buying processes and when each stage is likely to occur then you can anticipate what will be required so that you can prepare thoroughly for each stage in advance of it occurring. You will also then be in a better position to influence customers as they proceed through the buying process. The earlier you are involved, the more influence you have. For example, if you are involved at a very early stage then you may even be the trigger that initiates customers becoming aware that they have a need or problem in the first place.

That being said, it is more common for customers to have some idea that they have a problem or need. However, by being involved at a very early stage you are able to help them to explore the nature of their problem, help them to become fully cognizant of the costs or implications of the problem, and influence their thoughts on what a potential solution may look like.

If customers have a formal buying or procurement process (as is often the case with larger expenditure) then it is relatively easy to elicit the process that they will follow. If they do not have a considered (or even a conscious) buying process then it is helpful to discuss and agree with them the steps they will follow so that you can align to them, and adapt your selling approach as required.

Far too many salespeople firstly make premature proposals that are not aligned to where customers are in their buying process (ie they are made far too early) and secondly do not adapt their proposal to the specific needs, style and personality (more of this later!) of the customer. In addition, if you commence negotiation of commercial terms too early, you will have missed the opportunity to fully persuade customers of how suited to their needs your solution is and to maximize their perception of the value you will bring to them.

The specific interests and needs that the customer has in relation to the product or service in question

As mentioned above, it is vitally important to adapt all of your sales proposals so that they meet the identified needs of the customer in question. Although there may be many similarities between customers in the same industry or sector there will also be many differences. Good salespeople will make sure that they understand these differences. They will go to great lengths to understand the customer's situation in great detail. They will identify the problems, challenges and objectives of the customer. They will confirm and agree these and make sure that the customer is fully aware of the impact and cost of these challenges.

They will then show how the key benefits of their product or service help customers to solve their problems, remove their challenges and meet their specific objectives. Each sales presentation will be adapted to meet the specific needs and interests of each customer. In the modern world of selling, standard or 'canned' sales presentations have had their day.

In addition to the above, when selling to businesses it is important to bear in mind that people will have certain criteria, needs or interests depending upon the role that they fulfil in their organization and/or the role that they fulfil in the buying process. We can usefully think of five broad categories of buyer:

1. *Senior managers or executive buyers.* These buyers will be the senior leadership of the organization you are selling into. Examples include business owners, CEOs and managing directors.
2. *Technical buyers.* These buyers are individuals who have technical knowledge and expertise in the field or area of the product or service you are selling. Their specific job role will be dependent upon the nature of what you are selling. Examples include IT managers (for IT and technology sales), scientists or technical specialists (for industrial sales) and human resources personnel (for people development and training sales).
3. *Operational or functional buyers.* These buyers are people who will be involved in operational or functional aspects of their organization. Examples include supply chain, logistics or distribution managers, plant or factory managers, and sales managers.
4. *Legal buyers.* These buyers will be from your customer's legal function (or external legal support) and will be involved in legal aspects such as terms and conditions, risk exposure and liability.

5 *Procurement buyers*. These buyers are members of the organization's professional procurement department and are involved in the acquisition of the goods or services that the organization needs to operate.

Owing to the nature of their responsibilities, each of these five categories of buyer will have different interests, needs and requirements. Wise salespeople will adapt their selling approach according to these, so that they are communicating the benefits of their product or service in a way that is of most interest to these different buyers. For example:

- *Senior managers or executive buyers*. Owing to their senior position in the organization and their responsibility for making high-level decisions about strategic direction, strategic execution and policy these buyers will usually be more interested in and focused upon these areas. If you can demonstrate that your product or solution will help them to take the organization from where it is now to where they want it to be then you will capture their interest. It is important to understand their strategic priorities and strategic themes so that you can align your product and service to these. If you successfully do this then you can succeed in positioning yourself more as a strategic enabler, and perhaps strategic partner, than a transactional supplier.
- *Technical buyers*. Because of their technical specialism and focus these buyers will often ask the most in-depth and probing questions about the characteristics of your product or service. To have credibility with them, you must understand your product or service in depth. They will be interested in making sure it is 'fit for purpose' and in comparing it with their current understanding of what is available in the market. In some industries they will also have concerns about the compatibility of your product or service with whatever they are currently using. For example, they will want to be certain that your product or service won't interfere with the smooth running of what is currently being used, and that it will integrate with and preferably enhance it.
- *Operational or functional buyers*. These buyers will have an interest in how your product or service helps them with their specific functional or operational responsibility. These managers will be interested in anything that improves the capability, efficiency or effectiveness of their function. For example, a sales director will

usually be interested in growing sales, or a supply chain manager will usually be interested in improving the efficiency and effectiveness of the organization's logistics. I specifically use the words 'usually be interested in' deliberately. Although we may initially make some assumptions about likely areas of interest, it is dangerous to do so without checking. Different sales directors, for example, will have different areas of focus dependent upon the current situation with their sales force. By identifying what these areas of interest are, you can then adapt your sales approach to address these areas.

- *Legal buyers.* These buyers will usually be qualified legal professionals and will therefore have great interest in contractual matters, terms and conditions, the scope of the contract, areas of risk and liability and so on. It is good practice to engage with these buyers as early as possible in the sales process, so that you can understand and address any areas of concern sooner rather than later. Leaving such matters until later in the sales process can slow it down significantly. In many cases you will need to be accompanied by a legal professional from your organization, whom you will need to brief thoroughly so that he or she can be most effective in helping to bring the sales process to a positive conclusion.
- *Procurement buyers.* Many salespeople will report that procurement buyers are interested in only one thing: price! Whilst it is true that these buyers will often be the most price-focused of the five types of buyer, their interest is broader than this. They want to ensure that the products or services that they are buying meet the organization's requirements, for example in terms of quality, suitable quantity, within certain time periods, at certain locations and at the best possible price. Such buying decisions are rarely made on price alone (perhaps with the exception of certain widely available commodity items). Price will usually be a factor, but it is a mistake to assume it is the only factor. Despite what procurement buyers tell salespeople – it isn't just about price. Procurement buyers are also interested, for example, in value for money and suppliers who can reliably deliver what they need. It is worth noting that procurement buyers need to demonstrate that they have added value to the buying process by way of a saving or procurement improvement, and wise salespeople will adapt their sales approach to ensure that this is possible!

The personality and buying style of the customer

Every human being has similarities with every other human being. And every human brain has similarities with every other human brain. However, each of our brains is also totally unique – dependent upon factors such as inherited genetic factors and environmental influences.

As we shall see in Chapter 6 these differences are responsible for the different personalities we experience when we meet different customers. Some customers make very quick decisions, whereas some customers make very slow decisions. Some customers seem to want a huge amount of detail, whereas other customers seem to need only a brief summary. Some customers seem open and friendly, whereas others seem distant and cold.

Attempting to sell successfully to these different personalities can get challenging. Far too many salespeople assume (incorrectly) that the way they like to be sold to is the way their customers like to be sold to. Nothing could be further from the truth! One of the keys to sales success is to sell to all customers in the way that they like to be sold to, or rather sell to them in a manner that best suits the way that they prefer to buy. So a vital component of adaptive selling is adapting your own behaviour and communication to best suit the preferences of the customer you are selling to.

In Chapter 6 I will introduce you to a model of human behaviour that has been developed as a result of extensive neuroscience research. It will provide us with a powerful way to observe, identify and adapt our selling behaviour based upon the different customers we encounter and the personality profile they possess. This model will enable you to do three vital things:

- *Observe.* Sense and register the various personality cues from your customer's behaviour.
- *Classify.* Correctly interpret these and identify the customer's personality preference.
- *Respond.* Adapt your behaviour and selling style according to the preferences of the customer.

As we have seen, the effective use of adaptive selling depends on firstly selecting the correct sort of adaptation required and then employing the sales behaviours that properly correspond to this adaptation. There are six areas to be aware of and to take action on:

1. Elicit your customer's buying process and agree these stages with the customer.

2. Align your selling process to this buying process.
3. Adapt your selling approach accordingly, anticipating in advance what the customer will require.
4. Go 'upstream' in the buying process as much as possible to maximize your influence and to shape your customer's perception of what is required in a way that is beneficial to your solution.
5. Adapt how you articulate the benefits of your product or service to meet the specific needs, challenges, problems, interests and objectives of each and every customer.
6. Adapt your selling style to suit the personality preference of each customer.

So let us continue our journey into our customers' brains by discovering one of the most powerful models of human behaviour, which you can use to power yourself to new heights of sales success.

6
The *PRISM* model of human behaviour and adaptive selling

During two hot August days I was first introduced to a fascinating and cutting-edge profiling instrument that was the inspiration for this book. I was lucky to be attending a training programme to become a certified *PRISM* Brain Mapping practitioner. I was already familiar with a variety of traditional psychologically based psychometric instruments and had used them extensively in my work as a consultant and corporate trainer. I was in for a pleasant surprise, as *PRISM* was different to anything I had ever encountered before. Colin Wallace, from the Center for Applied Neuroscience, who has spent over 20 years studying human behaviour, was delivering the *PRISM* practitioner certification training, and he led those of us who were lucky enough to be attending on a fascinating journey through the incredible organ that is the human brain, the realms of neuroscience and the *PRISM* Brain Mapping instrument itself.

PRISM Brain Mapping is the world's most comprehensive online neuroscience-based behaviour profiling instrument. It can be used in a variety of ways – to help make better recruitment decisions, build better teams, develop leaders, improve performance, inform succession planning decisions, improve communication, improve customer service and the area that particularly caught my attention – *improve sales performance.*

ɟrove home from the first day of the *PRISM* practitioner training my ʋwn brain was working overtime! The exciting possibilities for improving salespeople's performance using the latest neuroscience research and the remarkable insights into human behaviour that the *PRISM* model offers sparked idea after idea in my brain! At about the halfway point on my journey home, heading north on the M1 motorway in the UK, the traffic ground to a halt and, as a result of a serious accident several miles ahead of me, remained stationary for the best part of two hours. Usually getting stuck in a traffic jam would be a source of frustration, but on this evening it gave me a lucky opportunity to grab my notepad and to start furiously writing my ideas down. The two-hour delay seemed to pass in an instant, but by then my initial ideas had begun to take shape.

After successfully qualifying as a *PRISM* Brain Mapping practitioner I approached *PRISM* and explained my idea for this book. They very generously offered their support and guidance – as well as a huge amount of very helpful information and research findings! Perhaps most importantly, they generously offered to provide each reader of *Neuro-Sell* the opportunity to complete the online *PRISM* Brain Mapping questionnaire and receive a copy of their own introductory *PRISM* Brain Map absolutely free of charge. If you visit **www.neuro-sell.com** you will be able to register, firstly to gain access to the online questionnaire that you need to complete in order to download your own introductory *PRISM* Brain Map report, and secondly to gain access to a series of exclusive resources that are available only to readers of this book.

The *PRISM* Brain Map report that you will receive will provide you with a fascinating insight into your personality and behavioural preferences. This will also help you to gain the maximum benefit from this book, so please take the time to visit and register at **www.neuro-sell.com** so that you can download your very own *PRISM* Brain Map report immediately.

PRISM represents a comprehensive synthesis of some of the latest thinking on how the human brain works and why people behave in very different ways. In contrast to traditional psychological models it is not based upon any one theoretical view of human personality, but is a combination of current knowledge of brain functional activity.

PRISM has been developed over a 15-year period to exploit some of the recent discoveries in neuroscience. Recent advances in neuroscience and brain imaging have enabled a more in-depth examination of the chemical, functional and structural aspects of the human brain and how it works.

The *PRISM* scales that measure each individual's expressed behavioural preferences were created and validated by Dr Tenday Viki. Dr Viki is a chartered psychologist, a university senior lecturer in forensic psychology, and a former visiting fellow at Stanford University.

At the root of the *PRISM* model is the fundamental fact that all behaviour is brain-driven. Brain development occurs as a complex and unique interplay between the environment into which a child is born and his or her genes. As a result of this each person has a unique way of perceiving and responding to the world that he or she lives in. Those recurring responses, which in part are inherited (nature) and in part learned (nurture), fall into patterns that can be referred to as behaviour preferences. All people will exhibit their own personal behavioural preferences through the things they say and do and also the manner in which they say and do these things.

PRISM theory is based on the following principles:

1. The brain is a dynamic, electrochemical system. No one part of the brain does solely one thing, and no one part of the brain acts alone. All our thoughts, emotions and actions are the results of many parts of the brain acting together to create patterns of activity.
2. In spite of the tremendous similarities between our brains, we all act differently and have unique abilities and distinct preferences, desires, hopes and fears. Although every human brain may appear to have a very similar structure and be organized in the same way, the key to each of our different personalities and behavioural preferences will be found in the fine-tuning of the neural systems and networks of each person's brain. These differences are what lead to the defining qualities of our personalities and behaviours.
3. One pivotal concept that underpins our understanding of the human brain and how it influences personality and behaviour is called neuroplasticity. This is the brain's ability to change its physical structure – the way the billions of neurons are connected together. These changes are the result of the various experiences we have (and do not have) and involve nature and nurture working together to create each unique brain. The brain never loses this ability to change and adapt. Neuroplasticity has replaced the formerly held position that the brain is a physiologically static organ, and allows us to explore and understand how the brain changes throughout life.
4. As described in Chapter 4, despite the commonly used metaphor, the brain is nothing like a computer! It does not process information or

assemble thoughts and feelings from individual bits of data such as strings of digits. Instead, the brain is largely composed of 'maps' – arrangements of neurons that represent entire 'chunks', concepts or templates of knowing, perceiving or cognition. Most cognitive functions involve the interaction of such maps from many different parts of the brain at once. The brain assembles perceptions by the simultaneous interaction of whole 'maps', relating one whole map to others and looking for similarities, differences or relationships between them.

The *PRISM* model of human behaviour encompasses three key interrelated factors that combine to generate human behaviour:

1. The brain's architecture, including the neural networks that operate within and between the brain lobes. As described in Chapter 3, the longitudinal fissure divides the brain into the left and right hemispheres and the central sulci and lateral sulci divide the frontal lobes from the parietal, occipital and temporal lobes. This structure is represented by the four quadrants of the *PRISM* model that are described below.
2. The level and flow of chemicals (ie neurotransmitters and hormones) within the brain. The *PRISM* model focuses on the effect of dopamine, oestrogen, testosterone and serotonin upon human behaviour, and this will be described below. A number of independent studies have confirmed links between these chemicals and the behavioural scales that you will see are contained within the *PRISM* model.
3. People exhibit their particular behavioural preferences. When they describe themselves they tend to choose words and phrases that emphasize behavioural characteristics they regard as central to who they are. Over time, these words become encoded in their speaking habits – including their facial expressions, tone of voice and body language.

In summary, *PRISM* is designed to explain behaviour in terms of the coordinated electrochemical activity that takes place within the brain's architecture. Fundamentally *PRISM* is about our attention to the world – how we see and respond to our environment and the people in it. It is about how we perceive and represent our environment.

On a broad level the brain lends itself to partitioning, based largely on its anatomy. All proposed divisions within the brain are, however, highly artificial and are created in response to the human need to separate things into neat, easily understandable units. However, we must always bear in mind that the brain functions as a whole and, with that caveat in mind, the *PRISM* model provides us with a useful schema that we can refer to when we are visualizing how our brains are organized. It is, therefore, based on scientific principles and facts that have been simplified into a workable model to facilitate understanding.

PRISM normally measures nine dimensions of human behaviour, but in this book we shall focus our attention on the four main quadrants of the model.

When you have completed the online *PRISM* questionnaire and downloaded your report you will see a circular image that represents the structure of the human brain (see Figure 6.1). The top half of the circle represents the front of the brain, the bottom half of the circle the rear of the brain. The circular image is also divided vertically into two sides. The vertical dividing line represents the corpus callosum, which you will recall from Chapter 3 divides the two hemispheres of the human brain. The left-hand side of the image represents the left hemisphere of the brain, and the right-hand side the right hemisphere of the brain.

FIGURE 6.1 How the brain is represented on the *PRISM* Brain Map

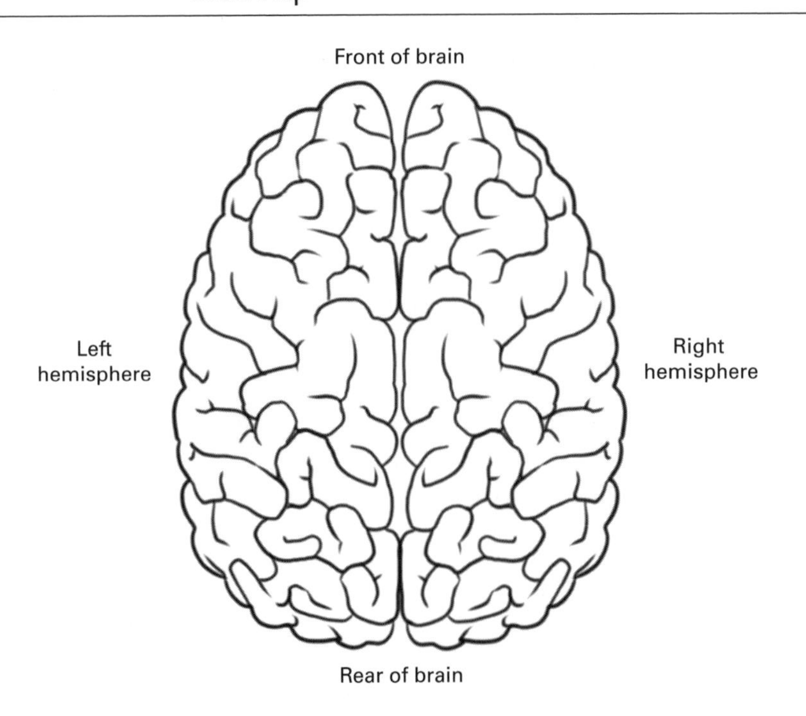

In addition, the circular image is divided horizontally. You will recall that the cerebral cortex is divided into four major lobes (frontal, parietal, temporal, occipital), and the horizontal dividing line represents the central sulci and the lateral fissure that divides the frontal lobes from the rear lobes of the brain. When combined with the vertical division, this produces four quadrants on the image that represent the four lobes of the brain (see Figure 6.2).

FIGURE 6.2 The four quadrants of the *PRISM* Brain Map

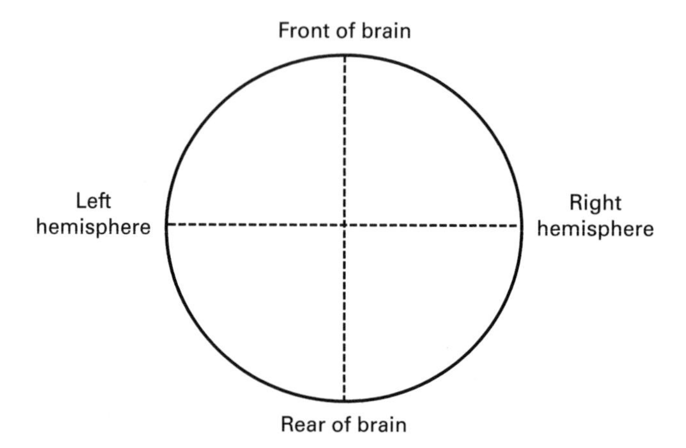

When you look at your own *PRISM* Brain Map you will see that each of these quadrants has a different colour – green, blue, red and gold (see Figure 6.3). The *PRISM* quadrant model is a graphic representation of how the brain's

FIGURE 6.3 The four colours of the *PRISM* Brain Map

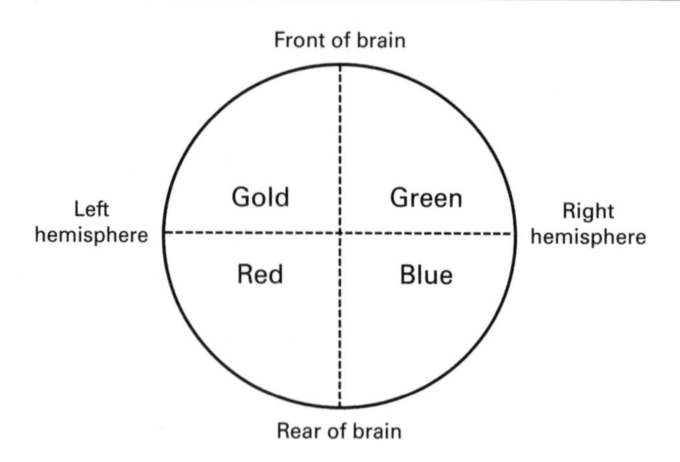

functional architecture and neural networks interact with brain chemicals to create behaviour, and each of the four quadrants is associated with specific behavioural preferences.

In short, the *PRISM* model summarizes how the brain creates behaviour and how specific behaviours are associated with brain areas. As you will see later, each of the four coloured quadrants is associated with specific behavioural preferences. When we understand these it will allow us to adapt our selling methodology and style to suit these behavioural preferences, thereby maximizing our chances of success.

Brain chemicals

As we have already explored the brain's architecture and how it is represented in the *PRISM* model, let us now explore some of the key brain chemicals and their influence and impact on behaviour. As with the incredible complexity of the architecture and structure of the brain, the subject of brain chemicals is also a complex area. What follows is a summary and simplification, which is designed to give you enough knowledge to understand how these chemicals contribute to human behaviour and, most importantly, how we can use this knowledge to sell more effectively!

Neurotransmitters

Neurotransmitters pass important information from one neuron to other neurons across a tiny gap (called the synaptic space or synaptic gap) between them and to other parts of the body, to orchestrate specific functions. Neurotransmitters (such as serotonin or dopamine) also produce the moods that affect our experiences. The brain chemistry we create on a daily basis, often by our own thoughts, determines how we feel.

Dopamine

Dopamine levels increase when the brain detects something new or novel or something that is unexpected. Dopamine release in the brain is also associated with activities that are pleasurable. For example, dopamine will provide us with a feeling of pleasure and happiness after we have eaten.

In addition to being connected with the experience of pleasure and happiness, dopamine is also generated by positive expectations or anticipations of experiences that the brain perceives as a reward – these include sex and positive social interactions. In this way dopamine can be thought of as the neurotransmitter of desire!

Elevated levels of dopamine in the brain produce focused attention, increased energy, motivation and goal-directed behaviours.

Noradrenaline

Noradrenaline is an excitatory chemical associated with physical and mental arousal, and heightened mood.

Serotonin

Serotonin has an effect on people's mood and behaviour control, conformity seeking, anxiety level and sleep. Drugs that alter serotonin levels are used to treat depression and anxiety disorders. Low serotonin activity is associated with aggressive, angry and impulsive behaviour, with high levels being associated with serenity and optimism.

Hormones

A hormone is a chemical released by a cell or a gland in one part of the body that sends out messages that affect cells in other parts of the body.

Many people have heard of the hormones testosterone and oestrogen. Neither of these hormones is actually made in the brain, but both can enter via the blood, and there are receptors in the brain that recognize them.

Testosterone is popularly referred to as a male hormone and oestrogen as a female hormone. In reality testosterone and oestrogen are produced and responded to by both men and women. It is the proportions that differ, with, for example, the male body producing more than 40 times more testosterone than the female body.

The environment in the womb during pregnancy can change the balance of the hormones that the foetus is exposed to, with some males being exposed to more oestrogen in their mother's womb, and some females to more testosterone. This can affect the degree to which the individual concerned expresses certain personality traits that are associated with testosterone and oestrogen. We will explore this later in the chapter.

When a person feels aroused or successful the cerebral cortex sends a signal to another part of the brain called the hypothalamus (an area at the base of the brain that regulates much of the body's hormonal activity) to stimulate the production of testosterone.

The brain's two hemispheres are connected by millions of nerve fibres, and individuals who were exposed to more oestrogen in the womb have more connections between these brain halves.

The four quadrants

To summarize, the *PRISM* model represents how the brain's functional architecture and neural networks interact with brain chemicals. Each of the *PRISM* model quadrants is associated with a region of the brain and with the influence of one or more neurotransmitters or hormones. As stated earlier, each of the four coloured quadrants is associated with specific behavioural preferences, and these reflect scientific studies into the effect of brain chemicals on behaviour.

Green quadrant

High noradrenaline and dopamine levels are associated with the following behaviours and are represented by the green quadrant on the *PRISM* model:

- unconventional;
- uninhibited;
- optimistic;
- mentally flexible;
- spontaneous;
- creative;
- impulsive;
- curious;
- opportunistic;
- energetic;
- reckless.

Blue quadrant

High levels of oestrogen are associated with the following behaviours and are represented by the blue quadrant on the *PRISM* model:

- caring;
- nurturing;

- sympathetic;
- idealistic;
- agreeable;
- supportive;
- sensitive;
- kind-hearted;
- charitable;
- tolerant;
- unassuming;
- sentimental.

Red quadrant

High levels of testosterone are associated with the following behaviours and are represented by the red quadrant on the *PRISM* model:

- competitive;
- independent;
- forthright;
- practical;
- aggressive;
- emotionally contained;
- decisive;
- direct;
- tough-minded;
- daring;
- focused.

Gold quadrant

High serotonin levels are associated with the following behaviours and are represented by the gold quadrant on the *PRISM* model:

- conscientious;
- conventional;
- cautious (but not fearful);

- detail-oriented;
- persistent;
- precise;
- orderly;
- consistent;
- careful planner;
- calm.

The four customer colours

So, with the above in mind, let me introduce you to four different colours of customer. Each of these customers thinks and behaves in distinct and different ways. Most importantly for us, their buying behaviour is different, and we need to understand this so that we can adapt our selling behaviour accordingly. If you understand how customers buy you can sell to them in a way that best suits them and will be the most comfortable to them:

- *The Green customer.* The Green customer will tend to be very curious and creative and will enjoy seeing things that are novel and new. In meetings Green customers will be energetic and may seem restless. If they are not absorbed in something that intrigues them they can get bored easily. They tend to be optimistic, spontaneous and autonomous. In the extreme, Greens can be hyperactive, impatient, impulsive and scatter-brained, and if you have a customer with very strong Green tendencies then as a salesperson you can add value to the relationship by providing balance to the impact of these tendencies. Greens love creating ideas and exploring what is possible. They will often think in metaphors and analogies, painting vivid pictures in their minds. Involving them in creating an innovative solution will help you to sell effectively to them. Since their preferred mental functioning is from the right brain, Greens are not bound by the limitations and mental barriers that restrict other behaviour preferences. They are possibility thinkers rather than probability thinkers. They rely on their hunches and insight to get a real feel for what is happening, and one of their greatest gifts is their highly developed intuition. Green customers will tend to be warm, innovative, optimistic, generous, fun-loving, adventurous, adaptable and trusting – sometimes unduly so. They get pleasure from variety and new experiences. If you can show them something new, novel or innovative then they will be very interested. Better still, involve them in creating an inventive solution. Then you will have their full attention!

- *The Blue customer.* The Blue customer will tend to be compassionate, warm, loyal and helpful. Blue customers will be polite, agreeable and tactful. They will take care to promote good feelings between themselves and the people they work with – they dislike conflict. They will try to smooth out differences and seek consensus. They tend to be consistent and reliable and maintain a steady pace. They are hard-working, patient and very productive. In the extreme, Blues can be hypersensitive, overly emotional, impractical, unrealistic and self-absorbed. As a salesperson if you can help the Blue to balance these areas then you will add value to the relationship. The quality of your relationship is something that is very important to this colour of customer. Blue customers need to connect with people at a deeper personal level than other colours of customer. This drive to connect with others is linked with oestrogen and oxytocin – a chemical produced, stored and triggered largely by oestrogen. When Blue customers make buying decisions they will consider the needs of everyone involved, as they want to please and to be valued. They like to create working environments where people will feel comfortable and valued. Their self-esteem is directly tied to, and influenced by, the quality of their relationships with others. Their emotional stability is based on how others react and interact with them. As buyers they are naturally cautious and approach any kind of change with trepidation until they fully understand the implications and the impact the change will have on other people whom they feel close to.
- *The Red customer.* The Red customer will tend to be very goal-oriented and will focus intensely on work. Red customers excel at what is called systemizing, which is the inclination to construct and analyse systems. People who are systemizers express more testosterone. Red customers will be assertive and authoritative. They can be very demanding and driven. They like to work in an environment that allows them to be in control and create results. In the extreme, Reds can become domineering, aggressive and pushy. They can be poor listeners and insensitive to others' feelings. The salesperson who can help them to balance their strong desire to take rapid action and help them to make effective decisions can add value

to the working relationship. As a result of their unusually focused minds, Red customers may display a tendency to be outspoken and blunt. Men and women who express high testosterone activity are, when under pressure, less likely to strive to be polite, respectful, considerate or friendly. They have little patience for anything that they perceive to be a waste of their time. They are irritated by long, repetitive discussions that do not lead to firm decisions allowing action to be taken. Reds are practical and pragmatic. They are direct and decisive and like to move with speed and efficiency. They are task- and action-focused.

- *The Gold customer.* Customers with a Gold behaviour preference will tend to be analytical, meticulous, guarded, rule-conscious, loyal, conscientious, thorough and competent. In the extreme, they can come across as arrogant, cynical, critical, distant and self-righteous. They believe in themselves, their capabilities and their intellectual abilities. To Gold customers the details matter. They will be process-oriented and will approach their buying decisions in a precise, consistent, sequential, linear, step-by-step manner. They have a strong preference for careful planning and for schedules and routines. They think concretely and are often literal, detail-oriented and orderly, as well as cautious. They are analytical thinkers with a strong need to be organized and accurate. They will want to determine the best way that something should be done, and once their plan is in place they will implement it without allowing themselves to be distracted until the task has been completed. Gold customers may speculate excessively about their buying decisions. They are slow to make important decisions because they are cautious and meticulous. At times they can suffer from 'paralysis by analysis', and the salesperson who can help them to make accurate buying decisions will add value to the relationship. Gold customers are driven by a need for perfection and want things to be done the right way. They want to make a careful, considered and correct buying decision the first time! They like to follow a process and dislike unpredictability or working at a fast pace, because of the increased risk of errors.

In this chapter we have explored how *PRISM* Brain Mapping looks at the structure and electrochemical activities of the brain, especially the interactions between the brain's hemispheres, in terms not only of attention and flexibility, but also of our attitudes to the world. It is about how we perceive and represent our environment.

We have seen that human behaviour is a combination of genetics (brain composition and sensitivity to certain neurotransmitters and hormones) and influences from our life experience. Through the infinite variables of genetics and experience each person develops different 'mental muscles'.

PRISM allows us to understand the different profiles or 'maps' that show, for example, how people prefer to process information. These preferred brain responses differ from person to person and form the basis of our observed behaviour.

The *PRISM* model offers four behaviour profiles for the astute sales professional to be aware of. These four types of people have different personalities and respond to the world (and make buying decisions!) in very different ways.

Although every customer will be a unique individual (no two people are alike), people will tend to have one quadrant that is what *PRISM* calls the 'tip of their behavioural spear'. This will be the individual's 'default' way of interacting with the world, and this should be the primary focus in the behavioural adaptation of the salesperson.

In Chapter 7 we will look at how the *PRISM* model will provide you with the ability to observe behaviour, draw sound conclusions about an individual's preferred thinking and behaviour and then adapt your selling approach and style to suit that of each customer.

7
How to read your customer and how to adapt your style

This chapter will provide you with a simple but effective methodology to:

1. *Observe.* Sense and register the various personality cues from your customer's behaviour.
2. *Classify.* Correctly interpret these and identify the customer's personality and behavioural preference.
3. *Adapt.* Adapt your behaviour and selling style according to the preferences of the customer.

The purpose of this adaptation is to minimize any interpersonal tension between you and the customer, in order to maximize your ability to sell successfully to that customer. As we will see later, the more you become like the customer, the more comfortable the customer will feel in your presence.

As described in Chapter 4, a fundamental organizing and operating principle of your brain that drives your thinking, behaviour and action is to avoid and move away from anything that is perceived to be painful, dangerous or threatening, and to move towards anything that is pleasurable, comforting or rewarding. Therefore the more comfortable the customer feels with you the less likely you are to be triggering the 'away from' response, and the more likely you are to trigger the 'towards' response.

The more primitive limbic system default setting is to be distrustful. It is more likely to initially judge a stranger as a foe than a friend. By using what you will learn in this chapter, you will be able to make a very powerful positive first impression with new customers that will provide a strong relationship foundation for the subsequent sale to be built upon.

As we have seen in Chapter 6, your customer's personality and behaviour are brain-driven. This chapter will enable you to observe external behavioural cues and from these make conclusions about what sort of brain is nestling inside your customer's head so that you sell to it most effectively.

Observe

I will start with examining customers' behaviour when you first meet them and, once this foundation has been established, move on to looking at some behavioural cues you may be able to identify before you make face-to-face contact.

In order to observe customers accurately you will need to focus on three areas:

1. their non-verbal behaviour;
2. their verbal behaviour;
3. their working environment.

These three areas will provide you with a series of cues from which you can draw conclusions on customers' preferences.

Behavioural cues

Table 7.1 outlines key behavioural cues to watch out for.

TABLE 7.1 Behavioural cues

	Non-verbal cues		Verbal cues	Working environment cues
	Energetic state	Body language		
Green	Energetic, outgoing, dynamic. Fast-paced. May move towards you quickly, be animated and shake hands enthusiastically on first meeting.	More open body posture. Big hand gestures and facial movements when talking. Pronounced facial expressions.	More verbally lengthy when talking. Can be loud, lively and enthusiastic. Personal or informal style with others – may tell stories and use humour. Great deal of tone and speed variation. May not listen well.	Modern office. Memorabilia of experiences displayed. Likely to be disorganized-looking, with lots of piles of paperwork.
Blue	Laid-back. Slow or steady. May walk towards you at a steady pace, greet you warmly and shake hands gently on first meeting.	More open, relaxed body posture. Slow, more contained hand gestures when talking. Intermittent eye contact – may look at the floor rather than you sometimes.	Conversational style, softly spoken. Warm, friendly, agreeable and sensitive. Steady and even tempered. Listens before talking.	Family and friend photos. Warm, homely environment. Loose system of personal organization.
Red	Energetic, impatient. Fast-paced, decisive. May march towards you purposefully, making firm direct eye contact, and shake hands very firmly on first meeting.	More contained body posture. Lots of frequent hand gestures when talking. Steady eye contact. May stand rather than sit.	Will get straight to the point. Little or no personal or small talk. Direct, forceful style of communication. Voice speed will increase when emphasizing urgency. Will increase voice volume to take charge. May interrupt.	Status or power symbols – large desk, trophies. Efficient but not necessarily particularly neat.

TABLE 7.1 *Continued*

	Non-verbal cues		Verbal cues	Working environment cues
	Energetic state	Body language		
Gold	Deliberate. May walk towards you at a steady and controlled pace, and shake hands briefly on first meeting. Slow or steady, step-by-step style.	More closed or tight body posture. Reserved movements, few or no gestures. Can appear expressionless. Will make reference to materials containing facts and data during meetings.	Little small talk. To the point, serious and precise. Discussion may be short or long depending upon the amount of information or data the person wishes to gather. Direct but calm and measured style; will question and clarify. Little voice inflection; may be monotone. May be periods of silence during conversation.	Functional office. Graphs and charts to track and control progress displayed. Very neat, tidy and organized.

Language cues

In addition to the cues in Table 7.1, the specific language and words that customers use and the questions that they ask will give you further cues to correctly classify them. Although sometimes they may very well use the specific words listed in Table 7.2, please also use these as general guidance for the areas and subjects of conversation that customers will be interested in and drawn towards.

TABLE 7.2 Language cues

Green language	Blue language	Red language	Gold language
adaptable	accommodate	achieve	accurate
alternatives	certainty	action	analyse
big picture	communication	bottom line	assess
creative	consider	competition	careful
exciting	cooperate	control	certain
experiment	coordinate	deadline	consider
flexible	empathy	decisive	data
freedom	feeling	deliver	detail
future	help	drive	facts
imagination	loyalty	efficient	figures
innovation	patient	fast	logical
inventive	people	focus	methodical
new	relationship-sensitive	goal	organized
novel	support	main points	plan
optimistic	teamwork	practical	process
original	trust	problem	quality
possible		results	sequence
see		solve	standards
variety		speed	step by step
vision		status	weigh up
		task	
		win	

As mentioned earlier, in addition to the behavioural cues that you can identify when first meeting and interacting with the customer, it is often also possible to gather some cues in advance of your first face-to-face encounter. You can do this by conducting some research in advance of meeting the customer. As we will explore in Chapter 8, sales professionals will always conduct research on new customers before meeting them for the first time – but more on this later! One of the easiest ways that you can research the customer is by using online resources. Depending upon customers' seniority and how active they or their organizations are online, a quick search may reveal information that can help you to anticipate what their behavioural profile is. Please note that I said 'anticipate' and not 'assume'.

It is good practice for a sales professional to anticipate a number of factors before meeting a customer. We can anticipate, for example, likely areas of interest or challenges that customers are experiencing dependent upon the industry that they are part of, possible concerns they may have about your product or service (based upon interactions with previous customers) or their likely behavioural preference based upon research that you have conducted. When you anticipate you prepare in advance based upon your current knowledge and understanding. This preparation can prove to be very useful.

However, you don't blindly assume that your current knowledge and understanding are correct. You will test it when you are with the customer. For example, you may ask, 'Mr Customer, several of our clients in the X sector have told us that profit margins are being squeezed as a result of increased overseas competition. Is that something that you are experiencing also?'

In the same way, you may anticipate the likely behavioural profile of customers from your online research, which will help you with your planning and preparation. But you will wait until you meet customers and interact with them before making your final decision about their behavioural profile and likely preferences.

The individual you are researching may feature in a number of ways online, for example being profiled or interviewed on their own or other websites. We need to be cautious, as sometimes this information will have been prepared by marketing or PR people or, in the case of interviews, may have been edited by a journalist, so we cannot always be sure that we are seeing an accurate reflection of the person's true personality and preferences.

One very powerful online research tool is the networking site LinkedIn. Increasing numbers of businesspeople are members of LinkedIn and will have a public profile that is available for you to view. The advantage of LinkedIn is that people complete their profile themselves, creating a description and using language that they are comfortable with. This can provide a rich source of clues that we can use to anticipate their behavioural profile.

By way of illustration here are some genuine descriptions from people I am connected to on LinkedIn. As I have worked with them I have a good insight into their personality and behavioural preferences, and it is interesting to see the cues their personal descriptions provide!

- *Green:*
  - '... enjoy finding and creating...';
  - 'I'm enthusiastic and passionate about...';
  - 'Developing a creative vision';
  - '... out-of-the-box thinker'.
- *Blue:*
  - 'I enjoy supporting and developing people';
  - 'building long-standing relationships';
  - '... coordinating people';
  - 'building teams and coordinating between departments'.
- *Red:*
  - '... close more business, faster';
  - '... get in control of business and drive excellence in execution';
  - '... dynamic and results-driven';
  - '... pragmatic, proactive and decisive'.
- *Gold:*
  - '... ensure adherence to... policy throughout the business';
  - '... considered approach to risk management';
  - '... strong attention to detail';
  - '... highly accurate'.

In addition to researching customers online you can gather evidence from early interactions with them via e-mail and over the telephone. For example, let us review the response we receive from each type of customer in response to an e-mail confirming an appointment. The e-mail you have sent says:

> Dear Steve
>
> Further to our telephone conversation, I am writing to confirm our meeting at your offices on Monday, 14 May at 2 pm.
>
> As discussed, it would be helpful to understand more about your new project and to discuss how our new service may be able to help you with this.
>
> If you have any questions or have any specific areas you wish to discuss at the meeting please let me know.
>
> I look forward to meeting you.
>
> Kind regards
> Simon

A Green response is:

Hi Simon

Great – time works for me!

Our new project is looking really exciting and I'm looking forward to hearing about how your new service can help us to innovate.

See you Monday – have a great weekend!

Kind regards
Steve

A Blue response is:

Hello Simon

Thank you for your confirmation. Please let me know if your plans change, as I am sure that I can move things around if needed.

Attached is a map and address details to help you find us. Please note that there are visitor parking spaces near the main entrance – this will be easier for you than the long walk from the main car park!

The proposed new project represents quite a significant change for the company, and I have some concerns about how people are feeling about it and how we can introduce it in a way that minimizes disruption to current ways of working.

I am looking forward to meeting you (and to learning about your people's capability) on Monday 14th.

I hope you have a pleasant and relaxing weekend.

Kind regards
Steve

A Red response is:

Simon

In the diary.
Need to know about your execution/resource capability.

Steve

A Gold response is:

Dear Simon

Thank you for confirming our meeting at 2 pm on Monday, 14 May at our Head Office in Meeting Room 4.

Attached is a map and address details. Please note the postcode you need to use for your sat nav is different from our main postal address postcode. The correct postcode for you is detailed on the bottom left-hand corner of the map.

Please sign in at reception. They will call me and I can come and meet you and take you to the meeting room. Because of our security procedure all visitors must be accompanied whilst on site.

The purpose of the meeting is to understand if you meet our criteria for a suitable external provider. If this is the case then we can shortlist you for the next stage of the tender process.

I will prepare to present:

1. Where this new project fits into our existing process.
2. Outline of the scope, stages and timeline of the new project.
3. Key deliverables in terms of quality, cost and time.
4. Key stakeholders and project participants from our side.

Please prepare to present:

1. Specific examples/case studies of where you have undertaken such work before.
2. Details of existing customers that we can use as references.
3. An indicative cost breakdown of your service provision.
4. Your internal quality control processes and procedures.
5. CVs of the consultants you would propose as being suitable for this project.

I would also be interested to understand more about your company, including your full product and service portfolio and profiles of your people's capability.

Please advise what information you can send in advance of the meeting on Monday, 14 May.

Yours sincerely
Steve

By attuning yourself to the cues in all of your customer's communication, you will maximize your chances of correctly identifying that customer's behavioural preference. This includes when speaking to the person over the telephone. We can often be in the situation where we will speak to our customers over the telephone before meeting them in person. And there may be people who spend the majority if not all of their time interacting with their customers over the telephone. If you refer to the verbal cues and language or word cues outlined in Tables 7.1 and 7.2 and start to pay attention to these, you will find that you can rapidly and successfully identify the customer's behavioural preference over the telephone too.

So we have observed the various personality cues from the customer's behaviour, and we now need to classify the customer.

Classify

From the information above you will start to be able to make an initial classification. Please remain open-minded as to the customer's colour for two reasons.

Firstly, some people will take a little longer to correctly classify, as their behaviour will not be so obvious or overt.

Secondly, as explained in Chapter 6 (and you will probably see this from your own *PRISM* Brain Map), most people have one quadrant or colour that is their main or primary preference ('the tip of their behavioural spear') and one that is a secondary or supporting preference ('the shaft of their behavioural spear'). Therefore, you will probably pick up cues from both (or perhaps even other) preferences.

As you become more confident and competent at classifying your customers you will start to consider finer definitions such as 'Blue/Gold' (main or lead preference Blue, supporting preference Gold) or 'Red/Green' (main or lead preference Red, supporting preference Green), and this will then allow you to adapt your behaviour even more effectively, as in addition to adapting to your customer's lead preference you can also incorporate some adaptations to the customer's supporting preference.

Now that you have made your initial classification you are in a position to adapt your behaviour accordingly to harmonize with the buyer's colour.

Adapt

Tables 7.3 to 7.6 outline the following for each colour preference:

1. what the customers like and are motivated by;
2. what they dislike;
3. the appropriate selling style and approach.

TABLE 7.3 Green customers

What they like and are motivated by	What they dislike	Selling style and approach
New ideas and ways of working. Innovation. Creativity. New experiences. Experimentation. Being an early adopter. Variety. Change. Thrills. Challenges. Approval and social recognition – being liked and popular. Expressing their ideas and opinions. Relating to people in a positive and friendly environment. A dynamic atmosphere. Freedom to be innovative. Doing things their way. As few rules as possible. Meeting new people.	Repetitive routine. Rigid processes and restrictions. Anything they perceive as ordinary or boring. Lots of details. Too much structure. Schedules. Only one option or answer. Narrow-mindedness. Slow pace. Being unpopular.	Present positive solutions. Present new ideas, novelty and innovations. Talk about the future. Involve them in generating and discussing ideas. Let them talk, show you are listening and smile to encourage them. Provide a big-picture overview and implementation timeline to future outcomes (keep detail in reserve). Use more emotion, less logic. Use big-picture colourful visuals. Provide a range of options and then help them to narrow their choice. Show how your product or service will make them stand out. Be friendly and enthusiastic. Tell stories. Have fun during your presentation. They can make quick decisions so be ready to move quickly and ask for the business or else your competitors may steal them!

TABLE 7.4 Blue customers

What they like and are motivated by	What they dislike	Selling style and approach
Maintaining stability and the status quo. Cooperation and collaboration. Consensus and harmony. Agreement. Loyalty. Friendly, low-stress atmosphere. Feeling part of a united team. Slow pace. Having time to make decisions. Helping other people. Putting other people first. Being valued and appreciated. Things that fit with existing ways of working, norms or culture.	Insensitivity. Rudeness. Aggression. Conflict. Change. Pressure. Impatience. Being rushed to make decisions. Unstable environment.	Invest time in social small talk. Be warm and friendly. Be prepared to share personal details. They will make slower decisions and may consult others to get consensus, so be patient. Offer to involve other people from their company in the process. Keep any changes or innovation steps small. Provide whatever is needed to show your product or service is safe and reliable. Don't push for a close – take it slow and easy! Move the sale forward slowly and steadily – a number of small sequential decisions will be easier for them than one big one! Take a collaborative, partnership approach of working together. Show how your solution helps and benefits others. In case studies show who you have worked with and how other people like them have benefited – use plenty of proof that it works. Be reassuring and stress the support they will receive. Let them speak to or meet other customers.

TABLE 7.5 Red customers

What they like and are motivated by	What they dislike	Selling style and approach
Winning. Taking action. Overcoming challenges and achieving goals. An environment that allows them to be in charge and achieve results. Deadlines. Achieving results. Being the best. Personal status and public recognition. Power and control. Leadership roles. Being in charge and having authority. Efficiency and productivity. Change (if they are in control). Making tough decisions.	Losing. Long, repetitive discussions. Slow pace. Indecision. Inefficiency. Anything they perceive as a waste of their time. Excessive detail and irrelevant information. Having to follow orders. Things that are too far away in the future that can't be started now.	Get straight down to business; be concise and businesslike. Keep meetings and presentations focused and short. Be prepared to be challenged and stand up to it to demonstrate credibility. Ask for their opinion – they are going to give it to you anyway! Use a key bullet point style – get to your point quickly and keep focused on the big picture. Focus on results, achievement and action. Show an overview of steps required to achieve results. Be practical and pragmatic. Show how your product or service will give them power, control and status. Case studies – focus on results. Show what you have achieved for similar customers. Show what you have achieved for previous clients; show how you have helped them to overcome challenges. Move fast; they will be comfortable if you push to move ahead. They make fast decisions, so be prepared to move quickly.

TABLE 7.6 Gold customers

What they like and are motivated by	What they dislike	Selling style and approach
Quality results. Accuracy. High standards. An environment that allows them privacy, peace and quiet with few interruptions. Logical, step-by-step approach. Predictability. Being methodical. Efficiency. Planning ahead in detail. Following process. Facts. Logic. Clarity. Time to complete things properly. Time to make the right decision. Getting things right. Tried and tested.	Carelessness. Disorganization. Lack of clarity and detail. Inaccurate information. Low standards. Incompetence. Emotional issues and displays of emotion. Being rushed. Change. Untidiness.	Be prepared and professional. Limit the 'small talk' – don't get personal. Take a slower, deliberate, step-by-step approach. Prove your capability and credentials – provide evidence. Provide whatever detail is needed to show the solution is tried and tested with proven results. Allow them time to ask as many questions as they can, and make sure you answer them succinctly and in detail. Use more logic, data, research and facts than emotion. Make case studies detailed and data-driven. Allow them time to consider, analyse and absorb information. Agree a process with them and follow it. Show data, information and studies, and leave additional detail with them. Give them time to analyse; don't pressure them into a decision or try to rush the deal.

Reference copies of the tables can be downloaded free of charge from the **www.neuro-sell.com** website once you have registered.

To finish this chapter, here are some phrases that you may like to use with each type of customer:

- *Green customer:*
  - 'This will enable you and your company to lead the way into the future.'
  - 'This will help you to take the lead in your market.'
  - 'This will put you at the cutting edge of your industry.'
  - 'This will provide you with the latest state-of-the-art/cutting-edge product/service/solution.'
  - 'I'd really welcome your ideas and input into how this might work.'
  - 'It would be great to be able to showcase you as a client who is leading the field.'
  - 'The main benefit to you is...'
  - 'This is the very latest solution/product/service.'
  - 'You are one of the first people to see this.'
  - 'Do you think you might have some other uses for this?'
- *Blue customer:*
  - 'If you would like to speak to our other clients who have made this transition successfully...'
  - 'This is a tried and tested and proven product that you can rely on.'
  - 'We are going to be around to help you whenever you need us to provide help and support.'
  - 'We'll make sure we take the time to carefully consider this with you before we go ahead.'
  - 'This will provide you with reliability and security.'
  - 'I will make sure I provide you with whatever information and reassurance you need to make the right decision.'
  - 'We can take some small baby steps with you.'
  - 'It would be helpful to gather the views of other people in your company whom this would help.'
  - 'Our guarantee is 100 per cent rock-solid and eliminates any risk on your part.'

- *Red customer:*
  - 'You're the type of person who would make this work.'
  - 'This will put you at the cutting edge and help you be a leader in your field.'
  - 'You'll be in total control of this.'
  - 'This will put you in the driving seat.'
  - 'This delivers results.'
  - 'Let me prove to you that this delivers results.'
  - 'This is a proven and effective way of...'
  - 'This will enable you to achieve...'
  - 'Let's agree the action steps needed to get started.'
  - 'We can move as fast as you want!'
- *Gold customer:*
  - 'Once you have taken the time to examine the facts...'
  - 'With the information I have provided you are in a position to examine the facts, interpret them and draw your own conclusions.'
  - 'I have brought along all of the information that you'll need to examine this product thoroughly and draw your own conclusions.'
  - 'This is a proven product, and our case studies demonstrate that this is something that you can rely on.'
  - 'We have achieved consistent results for Customer X. Would you like to have the details?'
  - 'We wouldn't want to go ahead until we were 100 per cent sure that it is right for you.'
  - 'Our focus on high standards...'
  - 'We constantly measure, assess and review our performance.'
  - 'Let me show you the details/facts/breakdown/research.'

You are now equipped with the ability to observe your customers, classify their behavioural preference and then adapt your selling style so that it suits their preference. This will lessen interpersonal tension and make the whole process more comfortable for everyone involved.

When I explain this methodology to my audiences as a speaker or to my clients when consulting with them, someone invariably asks if changing your behaviour in this way in some way makes you appear to be artificial or not yourself. If you have this concern then let me put your mind at rest. When you utilize this approach you do not suddenly adopt an entirely alien character to your normal way of behaving. You subtly and respectfully adapt. You become a little more Green, Blue, Red or Gold as appropriate. You flex and adapt your natural behaviour – you do not change it completely! You are still your authentic self, and this is important to making sure you come across with credibility and confidence.

Indeed, as you become more like the customer in subtle ways and the interpersonal tension disappears, the relationship between you improves and becomes more positive. Customers don't think anything is strange – they like you. You are like them! This allows you to sell to them in a way that is comfortable to them. And most importantly this allows them to buy in a way that is most comfortable to them. Sales professionals who can help their customers to feel comfortable when they are buying are the ones who are going to be the most successful.

As a closing thought, the chameleon never stops being a chameleon. It just changes its colour to adapt to and blend into its environment. And that is what this chapter has been all about – blending elegantly into the customer's preferred way of buying. So I would encourage you to become a chameleon and change your colour to match that of the buyer – to whatever degree feels comfortable to you. The only people who will dislike it are your competitors...

8
The 'Neuro-Sell' brain-friendly selling process – the first phase

Consider

The aim of the 'Neuro-Sell' brain-friendly selling process is to help customers reach a decision that is right for them. We all want to make good decisions. Our job as sales professionals is to serve our customers by helping them to make decisions that benefit them. The 'Neuro-Sell' brain-friendly selling process is customer-focused and customer-centric and actively involves customers in the decision-making process. Nothing is 'forced' or 'pushed' upon them.

It reminds me of a quotation that is attributed to the Chinese philosopher Lao Tsu: 'When the best leader's work is done the people say "We did it ourselves."' With apologies to Lao Tsu, my version is: 'When the best sales professional's work is done, the customer says "I've just made a great decision!"'

This process has not been dreamed up in an ivory tower. It is based on a powerful synthesis of cutting-edge neuroscience research, combined with a tried, tested and proven selling process that has been extensively field-tested in live selling situations. In short, this is a combination of cutting-edge science and hard-won experience!

There are eight brain-friendly stages to follow. Please note that these stages are intended to be a practical guide for you. There are no slavish sales scripts to follow or steps to be confined to. Think of them as a useful handrail to guide you through the sales process, from making a positive first impression all the way through to getting a firm commitment from the customer.

The eight stages are:

1. Consider.
2. Comfort part I: connect.
3. Comfort part II: chameleon.
4. Comfort part III: control.
5. Context and catalyse.
6. Check:
   - cash;
   - criteria;
   - authority;
   - pain;
   - pleasure.
7. Convince:
   - curiosity;
   - clarity;
   - contrast;
   - concrete;
   - certainty and credibility.
8. Confirm and conclude.

For ease of reading and understanding I have divided these eight brain-friendly selling stages into five phases, with each phase occupying a separate chapter. The first phase concerns the planning and preparation you need to do prior to meeting the customer. The second phase is devoted to maximizing the customer's sense of psychological comfort. The third phase concerns gathering and clarifying information on the customer's situation so that you can understand what the customer needs. The fourth phase is devoted to presenting or pitching your products and services effectively. The fifth and final phase is focused on closing the deal and winning the business.

So let us start our journey with looking at what you need to do before you meet the customer to set yourself up for maximum sales success.

Stage 1: consider

The first stage of the 'Neuro-Sell' brain-friendly selling process is to fully and properly *consider* the customer and the sales call you are about to undertake. As mentioned in Chapter 1, in business-to-business sales the amount of time that buyers are willing to give to sales professionals is reducing. It is getting harder to get in front of buyers, and therefore when you do get in front of the customer it is vitally important that you are well planned and prepared.

If you are selling direct to consumers, as markets become more and more competitive the number of companies vying for consumers' attention and money will increase. Getting in front of consumers will become more challenging. It is vitally important that you maximize your ratio of meetings to closed sales, so once more being planned and prepared is vitally important.

As all true sales professionals know, it is important to spend time considering what you need to plan and prepare in advance in order to maximize your chance of success. So let's consider some of the vital elements.

Information on the customer

If it is new customers that you are visiting, it is important to research them prior to meeting them. The availability of information available via the internet means that research can be conducted quickly and easily.

If you are selling to businesses, in the vast majority of cases they will have a website that you can review, and increasingly businesses will have a social media presence. Conducting a quick but thorough review will usually provide you with valuable information on the nature of the customer's business.

It is increasingly becoming expected by business customers that you will have researched their company prior to you meeting them. When you have done this you can refer to what you have learned about their business and ask further questions to understand more (we will look at the sales questioning process in greater detail later in this chapter). Showing an interest in their business is also a good way of establishing rapport with customers.

Information on the customer's industry

If you are selling to specific industries on a regular basis, it is also important to keep up to date with the latest developments in these industries. You may wish to subscribe to relevant trade magazines, visit industry-related websites or subscribe to e-mail newsletters and newsfeeds. This will keep you abreast of what is happening, provide insights into the problems and challenges the industry sector is facing and provide useful conversation topics when you are meeting your customer. The ability to discuss the latest news or trends knowledgeably with your customers will help to position you as an authoritative sales professional.

Research by Dr Robert Cialdini (1993) from Arizona State University shows that people feel a strong pressure to comply with requests from an individual whom they perceive to be an authority. Society socializes people that obedience to authority is the correct way to behave. Cialdini states that 'it is frequently adaptive to obey the dictates of genuine authorities because such individuals usually possess high levels of knowledge, wisdom and power'.

Therefore it is important that you are perceived as an authority and expert in your field by the customer. Customers are far more likely to respond positively to your request if they perceive you to be an authority. Hence there is a need for you to be knowledgeable and keep up to date in your chosen field so that you are able to communicate your authority to the customer.

We must always be focusing on making the customer's brain feel as comfortable with us as is possible. If you are perceived as an authoritative expert then the customer's brain will feel safe in your hands and be more open and receptive to your advice and suggestions.

Cialdini's research shows that deference to authority can occur in a 'mindless fashion as a kind of decision making shortcut'. Therefore it is vitally important that you work on establishing and communicating your authority and expertise.

Information on your products and services

In addition to researching the customer's business and being well informed on the customer's industry, you have to be able to demonstrate your expertise in the products or services you are selling. You must know your products and services in detail and be able to answer with confidence and clarity almost any question a customer might ask.

If the customer's brain senses any unease on your part (for example, as a result of a lack of product knowledge, or your inability succinctly to answer a question), then the customer in turn will start to feel uneasy. You must exude confidence and authority, as this will make the customer's brain feel comfortable and make the customer receptive to your suggestions.

Make sure that you are thoroughly planned and prepared and have whatever materials, research, literature and other sales collateral you need with you. The impression you must create in the customer's brain is: 'This person really knows his/her stuff!'

Meeting goals

The amount of research that has been done into goal setting and its impact on performance is impressive. One of the most robust conclusions to come from all of this research is that goal setting improves task performance.

As I state in my book *The Inner Winner* (Hazeldine, 2012):

> Goals influence performance in a number of important ways:
>
> - Goals focus attention and action on important aspects of performance
> - Goals set specific standards that motivate individuals to take action
> - Goals increase not only immediate effort and intensity, but also help to prolong effort and increase persistence
> - Goals also prompt the development of new problem-solving and learning strategies.

In short, setting a clear goal (or goals) for your meeting gives your brain something to focus and lock on to. On a day-to-day basis our brains are bombarded with sensory information and, in order to manage this, certain information is 'filtered out'. The part of the brain called the reticular activating system, to which you were introduced in Chapter 3, is what decides which information it is important to pay attention to and what can be ignored. It helps the brain to decide what to consciously focus attention on.

People who live close to airports or railways are not as aware of the noise of the planes and trains, as their reticular activating systems dampen down the effect of the repeated stimuli. This helps to prevent the brain being overloaded. By contrast, if you have ever become interested in purchasing a particular model or colour of car, because the goal that you now have in your brain influences what your reticular activating system pays attention to, the world will seem to be full of exactly that model and colour of car!

Therefore, to maximize your chances of sales success it is very important to have clarity about the goals that you have for your meeting. Fundamentally, every customer meeting should have one of two goals: 1) to close the sale; 2) to advance the sale towards a close.

In many industries, particularly where the sale is more complex or where the level of expenditure is high, it is not practical to close the sale in one meeting. It will take several meetings for the sales process to be concluded. In these cases you need to have a clear goal to advance the sale. This will prevent unproductive meetings and the sales cycle being lengthened more than is absolutely necessary.

Your ultimate goal is to close the sale. This goal can be broken down into several sub-goals that will support this. The sub-goals could include information you need to gather in order to move the sales forward, understanding the buying process that will be followed, identifying who will be involved in the buying process, determining what it is you want the customer to believe about you and your organization so that the customer feels comfortable about working with you, and so forth. It is important to make these as clear as possible so that you have clarity over what the goal is. Without some clear evidence you cannot know if you have been successful or not.

Poorly articulated objectives such as 'Build the relationship with the customer' or 'Keep in touch with the customer' will deliver poor meeting results and waste your time and, most importantly, the customer's time. Make your meeting goals very specific. Give your brain something to lock on to.

Here are some examples:

- 'At the end of this meeting I will be able to articulate the nature of the customer's specific challenges and problems and gain the customer's agreement to these.'
- 'At the end of this meeting I will be able to articulate the financial cost of the customer's specific challenges and problems in both the short and the long term and gain the customer's agreement to these.'

- 'At the end of this meeting I will have identified the steps in the customer's buying process and which people will be involved at each stage, and will have gained the customer's agreement to these.'
- 'At the end of this meeting I will have defined the criteria that the customer will use to determine which supplier the customer wishes to place the business with and will be able to gain the customer's agreement to these.'
- 'At the end of this meeting I will gain the customer's agreement that we have the experience, expertise and capability to be the customer's new supplier.'

By having a very clear goal for the meeting, with evidence that you can use to determine if you have met your goal, you give your brain something very concrete to lock on to. You will have opened the specific sensory filters that will support you in achieving your meeting goal, making your brain more attuned to noticing the information that it needs to locate. The simple but powerful step of defining the specific goal or goals you want to achieve in every meeting will powerfully support your sales success.

Researching the person or people you will meet

In addition, as mentioned in Chapter 7, researching the individual people you will be meeting using LinkedIn is a valuable exercise. Along with the behavioural cues that will allow you to consider and prepare to adapt to their preferred behavioural style, you can also learn about their career history, identify areas of professional interest from their membership of LinkedIn groups and find out the sort of people they are connected to.

Arriving at the customer's premises

When you arrive at the customer's premises, keep your eyes and ears open for information that might be helpful. You can sometimes learn a lot from a customer's foyer from copies of internal magazines, annual reports, marketing material, leaflets, posters, pictures and awards on display. You never know when information you find will be useful to you. So don't waste your time sitting down drinking coffee – take a good look around! There is another very important reason never to sit down in a customer's foyer, and I will come to that later in the book.

Be ready to adapt

As mentioned in Chapter 7, you need to consider carefully what you have discovered so far about the customer's possible behavioural preference.

It is good practice to prepare to meet any possible behavioural preference (or combination of preferences) that you may come across. Be ready to meet and sell to any possible combination of Green, Blue, Red or Gold customers. For example, be ready to go into detail for the Gold customer, be ready to focus on action with the Red customer, and so forth.

That being said, if your research gives you the indication that the person you are meeting is very likely to be, for example, a Blue personality preference then you can consider this and prepare accordingly. However, always be prepared for the customer's personality to be different from what you initially expected and for other people to join the meeting. Always expect the unexpected!

'Neuro-Sell' Pre-Call Planner

When you visit and register with the **www.neuro-sell.com** website you will be able to download a free-of-charge copy of the 'Neuro-Sell' Pre-Call Planner document. This document will take you step by step through effective pre-call planning. By using this document before *every* customer meeting you will be maximizing your chances of success.

Now that you have thoroughly considered all of the planning and preparation required to be a success when you meet the customer, you have established a solid foundation upon which you can build a successful sales visit. In Chapter 9 we will move on to meeting the customer and making the customer feel comfortable with us.

9
The 'Neuro-Sell' brain-friendly selling process – the second phase

Maximize comfort

In this chapter we will look at the second phase, consisting of stages 2, 3 and 4 of the 'Neuro-Sell' brain-friendly selling process, which are all focused upon maximizing the customer's sense of psychological comfort.

Stage 2: comfort part I: connect

You will recall from Chapter 3 that, when you first meet the customer, you are a stranger, and the more primitive parts of the customer's brain will instantly conduct a threat response and decide if you are friend or foe. Please don't be offended by this – we are talking about a mechanical, survival-orientated, selfish and unconscious part of the customer's brain.

In addition, as a result of the negative stereotypes that exist about sales-people, customers may be concerned that the salesperson they are about to see might try to pressure them to buy, trick or con them in some way, over-promise and then under-deliver, over-inflate the price of the product or service or sell them something they don't need or far more than they need. Customers could be feeling quite uncertain and insecure, so it is very important that at a very early stage of the sales call we maximize customers' sense of comfort and then not only maintain this but increase it during the entire sales process.

There are some specific things we need to do to calm the reptilian and emotional parts of the customer's brain and make it feel comfortable with us when we first connect with the customer.

Based upon what you learned in Chapter 7, as soon as you see customers you will start observing their behaviour so that you can successfully classify their behavioural preference and respond accordingly. This is designed to maximize your ability to make customers feel comfortable with you and enhance your ability to connect with them.

However, there are some 'universal' behaviours that you can exhibit that will help a customer's primitive brain to feel safe rather than threatened, helping it to calm down and begin to feel comfortable. These are:

- *Smile.* I appreciate that this may sound obvious, but a smile sends a strong signal of friendliness and acceptance to the other person that the person's brain will respond to positively.
- *Use open, relaxed body language.* This communicates a non-threatening manner to the customer's primitive brain.
- *Make sure your torso is facing towards the customer.* When we like someone we tend to turn towards that person (the opposite extreme is turning your back on someone!). People will perceive us as being more open and honest when they can see our torso.
- *'Flash' your eyebrows at the customer when you first make eye contact.* This is a universal body language signal where you raise your eyebrows for about a sixth of a second when you are about to make social contact with someone. This gesture communicates that you are feeling positive about meeting the person.
- *Keep your voice modulation and tone calm; keep your voice speed controlled and gentle.* Our voice mirrors our emotional state, so your voice needs to communicate a calm and non-threatening emotional state.

- *Mirror the handshake pressure of the other person.* As a default make your handshake firm but not too strong.
- *Don't invade the customer's 'personal space'.* Once you shake hands step slightly backwards and to the side.

In terms of the importance of personal space, a researcher called Edward Hall (1998) coined the phrase 'proximics', and this is a sub-category of the study of non-verbal behaviour, which we will explore in greater detail in Chapter 14. Hall discovered that all animals, including the human animal, need a certain amount of space to feel safe. There is a cultural element (some cultures demonstrate differences in how close people can be to each other in various situations), and there is a limbic element. If you violate someone's socially acceptable space by, for example, sitting or standing too close you will trigger a negative limbic reaction and heighten that person's brain's threat response.

Hall defined four distances and, although he did not mean these measurements to be strict guidelines that translate precisely to human behaviour, they are useful rules of thumb to follow:

- Intimate distance is reserved for interactions with lovers, children, close family members and close friends. This starts at touching and extends to about 46 centimetres (18 inches) apart.
- Personal distance for interactions among good friends or family members starts around 46 centimetres (18 inches) apart from the person and ends about 122 centimetres (4 feet) away.
- Social distance for interactions with strangers, newly formed groups and new acquaintances ranges from 1.2 metres (4 feet) to 2.4 metres (8 feet) away from the person.
- Public distance includes anything more than 2.4 metres (8 feet) away, and is used for speeches and lectures. Public distance is essentially that range reserved for larger audiences.

Do not get too close too early is the rule, as this will provoke the negative limbic response described earlier!

If you get this initial contact correct you will have successfully started to minimize your perceived threat response and increase the customer's level of comfort with you.

Stage 3: comfort part II: chameleon

As we discussed in the last chapter, it is vitally important to become a behavioural chameleon and adapt your behaviour to best suit that of the customer. In addition to this, and to the universal behaviours described above, there is another key universal behaviour that you can adopt that will positively contribute to lessening any interpersonal tension and making the customer feel comfortable with you.

Postural echoing

Research into non-verbal communication (Condon and Ogston, 1966; Kendon, 1970) reveals that when two (or more) people feel comfortable in each other's presence (for example, people who are friends) they unconsciously adopt similar body postures. The more friendly and in agreement they are, then the closer what is known as 'postural echoing' becomes. In addition they will display 'gestural echoes', synchronizing movements such as leaning forwards, crossing and uncrossing legs, nodding in agreement, and picking up their glasses to drink at the same time. Research using slow-motion film has shown that a 'micro-synchrony' of very small movements, which it is almost impossible to see with the naked eye, also occur. People who are feeling comfortable with each other will also echo each other's vocal qualities, speaking at a similar volume, pace and style. This behaviour is usually seen in people who perceive that they have the same status, that is that they are at the same level.

This phenomenon is believed to be a non-learned and instinctive human behaviour, which you may also hear referred to as 'isopraxis'. Small babies begin to mimic the facial expressions of people at a very early age in what is believed to be a survival-driven bonding activity.

You will recall from Chapter 3 that Professor Iacoboni, who conducted research into mirror neurons, believes that mirror neurons send messages to our limbic system and enable us to tune into, empathize with and connect with each other's feelings. Mirror neurons may explain what many psychologists have believed for a long time, that when you mirror another person's posture, gestures, movements, voice tone, voice pace and voice pitch you build a sense of comfort and rapport with that person at an unconscious level.

Many studies have shown that humans communicate more information through body language and voice tone than they do through the words used, and hence the reason that mirroring the customer in this way contributes powerfully to the sense of comfort and connection the person feels with you when you do it.

The existence of mirror neurons may also explain why so much of human communication occurs in this way. Your customer's brain is constantly mirroring your posture, movements, voice pace and pitch and also your emotions. This ability allows us to establish a deeper sense of mutual liking, understanding and comfort. It helps us to come to a shared attitude and agreement about how we can work together.

The practical application of this for you as a sales professional is to consciously begin the process of mirroring or echoing the customer as soon as possible. This will send a strong unconscious message of familiarity, liking and rapport. It is important to do this subtly and respectfully, gradually becoming more and more like customers in posture, gesture, movements, voice pace, tone and speed. As you become like them they will increasingly like you.

This behaviour will help you to make them feel comfortable with and receptive to you. You can then build upon this universal echoing by using your observations about your customer's behavioural preference and adapting your sales approach as outlined in previous chapters accordingly. As Dr Robert Cialdini from Arizona State University states in his book *Influence* (1993), 'People prefer to say yes to individuals they know and like.'

The more you are like customers from a behavioural perspective the more they will like you. The more they like you the more comfortable they will feel. The more comfortable they feel the more likely they are to spend time with you, share information with you and ultimately do business with you.

Stage 4: comfort part III: control

As you are moving through the early stages of the sales process it is important to keep focused on maximizing the customer's sense of comfort with you. Your aim is to continually contribute to creating a state of mind where the customer's brain is open and receptive to your sales message. The more comfortable the customer feels about you, the more open the message receptors in the customer's brain will be.

At this stage of the sales process you can add to customers' sense of comfort by providing them with a sense of control over what will happen. You can do this by outlining the process you are going to follow. Explain that you won't be launching into any sort of pre-prepared sales pitch. Explain that, firstly, you would like to understand about them, their business (if selling business to business) and their goals. Stress that it is only once you understand their situation that you will be in a position to see if you can help them or not.

When selling to new clients I always use a phrase such as 'if I can help you or not'; for example, 'If I think I can help you – and I do mean if – then I would like to make you a proposal.' I do this for several reasons. Firstly, it reduces any concerns that customers may have that, like far too many salespeople, I am going to launch into a canned sales pitch. I am only going to make them a proposal if I can help. Secondly, it suggests that I am selective about who I work with (which is true), and this helps to establish my credibility. It also introduces a sense of scarcity in that I may decide not to work with the customer. Dr Robert Cialdini's research also discovered that scarcity was a powerful principle of influence.

Things that are difficult to attain or obtain are perceived as more valuable and desirable. At an unconscious and more primitive level of the brain scarcity can be perceived as a threat to survival and arouses powerful emotions that can overwhelm the conscious or rational mind. I have found that introducing the slightest suggestion or possibility of scarcity makes the customer want my services all the more. For example, when I am contacted by conference organizers to discuss booking me as a keynote speaker, I will always ask them to tell me about their organization and what they want to get from the conference. I explain that 'I want to see if I am the right speaker for you, or not.' The phrase 'or not' positions me as a speaker who is not desperate to get the speaking engagement, that I won't work for just anyone, and that I will speak at a conference only where I will add value to the aims of the conference.

When meeting customers for the first time I also stress that they will always be the ones in control, for example by using language like 'You will be the one to decide if we are the right partner for you.' This acknowledgement that the customers will be in control will be perceived as rewarding by their brain. It also gives them a status boost, which again will be rewarding to their brain.

It is important not to overdo this. You want to position yourself as someone of equal status, and an authority in their field. You do not want to be perceived as subordinate to customers, as this will have impact later in the sales process – particularly during the negotiation phase! You can exude calm confidence by mentioning that you appreciate that they have a choice over who to select as a supplier and you are interested to talk with them to understand their situation and to see if you can be of help to them, or not. This calm confidence will help you to exude an air of authority and certainty that will be very appealing and comforting to their brain.

Another advantage of outlining the process that you will follow is that it increases customers' sense of certainty. The brain likes to try to predict what will happen. In the primitive world of 100,000 years ago, the ability to predict successfully what would happen maximized your chances of survival. Certainty is a rewarding experience for the brain. Uncertainty on the other hand arouses the limbic system in a negative manner. Until you outline what will happen and when, there may be a degree of tension or discomfort on the part of customers. Making them comfortable and certain about what will happen will help to get them into an open and receptive state of mind.

As a result of the way the brain operates, and the fact that it is more likely that the 'away from' threat response will be activated, we have to focus actively on increasing feelings of comfort, certainty and reward throughout the sales process.

So, by now, we should have a customer who is feeling comfortable with us and who will be open to us gathering information so that we can be of maximum help to the customer. In Chapter 10 we will explore how to do exactly that.

10
The 'Neuro-Sell' brain-friendly selling process – the third phase

Establish context and catalyse

The one factor that differentiates true sales professionals from the less capable salespeople is their ability to develop an in-depth understanding of their customer's situation and needs. This chapter focuses on doing this to an advanced level and covers stages 5 and 6 of the 'Neuro-Sell' brain-friendly selling process.

Stage 5: context and catalyse

This stage of the sales process is about gaining an in-depth understanding of customers, the context or circumstances of their situation, and their needs, goals, challenges and problems. Once you have done this you can then use what you have discovered as a catalyst to motivate customers to take action.

Just as medical doctors will not prescribe treatment until they have diagnosed a patient's symptoms, so it is vitally important to diagnose a customer's problems before offering any form of solution. However, with the 'Neuro-Sell' brain-friendly selling process we are going to do more than gain an understanding for ourselves. We are going to help and consult with customers so that they come to their own conclusions and insights about what they need to do that is in their best interests.

The four types of question

As described in my first book, *Bare Knuckle Selling* (Hazeldine, 2011b), broadly speaking there are four types of questions that you can use in a selling situation. These are:

- closed;
- open;
- probing;
- summarizing.

These can be remembered by way of the mnemonic COPS.

Closed questions

These are used to obtain a specific answer and to check facts. Examples include:

- 'Was it a success?'
- 'Is that the most important area?'
- 'Does anyone else need to approve this purchase?'

Closed questions usually result in a 'yes' or 'no' answer.

Open questions

These are broad, diagnostic questions that encourage customers to talk about their circumstances. Open questions usually start with words such as 'what', 'when', 'why', 'how', 'where', 'who' or 'which' and usually result in a multi-word or sentence answer. Examples include:

- 'What do you want to change or improve about your business?'
- 'What could your current supplier do better?'
- 'How do you currently handle customer queries?'
- 'Why are you considering a new supplier?'

It is important to stress that, whilst open questions usually result in a multi-word or sentence answer and closed questions usually result in a single-word answer, it is not always the case. Sometimes you will get a 'yes' or 'no' to a good open question and a long reply to a closed question!

Although many salespeople have been taught to ask open questions rather than closed questions (as this helps with information gathering), both open and closed questions are important and have their place in the selling process. Open questions are used to gather information and closed questions are used to clarify what you discover and get specific answers and commitments.

Probing questions

These are used to explore a point a customer has made. They allow you to drill further into what the customer has said so that you can understand it in more detail. This is sometimes called 'chunking down' and is a concept that we will return to later. Examples include:

- 'When you say you need to move quickly, how quickly do you mean?'
- 'What makes you say that?'
- 'In what way do you think...?'
- 'Give me an example of...?'
- 'How do you mean?'
- 'Why did you mention that particular feature?'

A useful probing technique is to use 'echo questions'. An echo question is where you use the last word or few words of what the customer says as a probing question. For example, if the customer says 'We need a supplier who is reliable', the echo question is 'Reliable?' In this example, you are probing further to discover how the customer defines 'reliable'. If you did not probe, you could make some assumptions about what 'reliable' means to you.

It is important to 'chunk down' or 'decode' the language that the customer uses. This ensures that you understand *exactly* what the customer means by 'reliable'.

Your customers will have criteria that they use to decide if a supplier is reliable. Some of these they will be conscious of and some unconscious. The criteria can form part of the 95 per cent of human cognition that is unconscious. By diving deeper into customers' language and therefore their thinking (both conscious and unconscious) you gain a more accurate and precise definition of the criteria that they will use to make a decision.

Asking probing questions around the customer's buying criteria will allow you firstly to identify what they are and secondly to prioritize them. Useful questions include:

- 'What is important to you?'
- 'What is most important to you?'
- 'The last time you took a decision like this, how did you decide?'
- 'What criteria will you be using to make your decision about which supplier to award the contract to?'
- 'Of the four criteria you mentioned, which of these is the most important to you?'

In addition to helping you understand customers' criteria, you may also help them to become consciously aware of the elements of their thinking – in this case their decision-making criteria – that were previously unconscious. This can help them to make a better decision and, as you were present when it happened, this insight will often be attributed to you, which will help your credibility, connection and relationship with the customer.

Summarizing questions

These are used to sum up the conversation you have had with the customer and to confirm the discussion you have had so far. This helps to keep the sales call on track and to check and clarify your understanding. Examples include: 'So if I understand correctly, is what you are saying...?' and 'So have we agreed that...?'

A very elegant method of really understanding what your customer wants, needs and values is to combine questions in a questioning funnel. You start with broad information at the top of the funnel, and using a combination of open, probing, summarizing and closed questions you get very specific information at the bottom.

At the broad top of the funnel you ask more open questions that encourage customers to tell you about their circumstances and what is important to them. You then use probing questions to dig deeper, chunking down to gather more information about specific areas. You can then use closed questions to clarify information and check specific facts, and summarizing questions to wrap up the questioning process. You can use a series of funnels

FIGURE 10.1 The questioning funnel

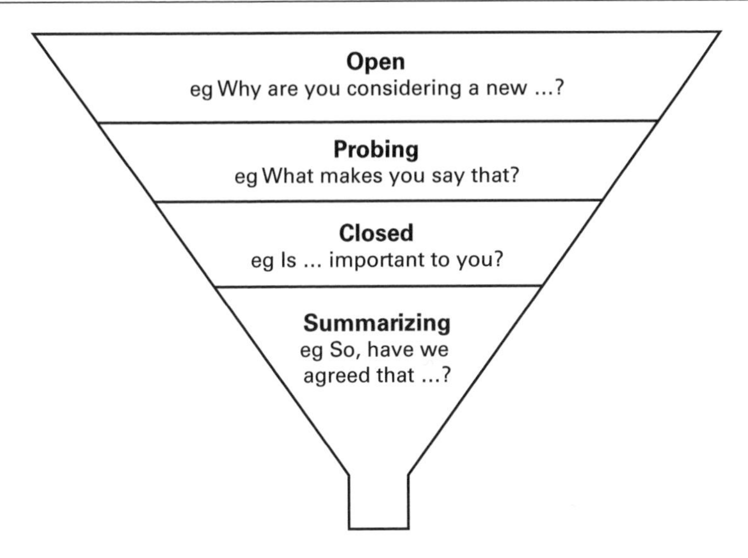

to gather all of the information you need to fully understand customers' needs, wants and criteria.

I sometimes liken this process of questioning to gathering together the pieces of a jigsaw puzzle. Imagine that you wanted to complete a jigsaw puzzle but did not possess a picture of what the completed puzzle looked like. In that case it would only be once you had all of the pieces together that you would be able to see the full picture.

In the sales process it can be dangerous to assume that you know what the full picture is, because that means you stop asking questions! You have to build a full and complete picture of customers' circumstances and the problems they face before you can build a picture of the solution with them.

Using chunking-down questions

I referred to the concept of chunking down when describing probing questions. It is useful to think of this concept in a broader way than just asking probing questions.

As the main aim of the 'Neuro-Sell' brain-friendly selling process is to help customers come to a conclusion about the best course of action to take, the more of their unconscious thinking processes (including their neural maps) that they are conscious of, the more of an informed decision they will be able to make.

You will recall from Chapter 4 that these 'maps' or arrangements of neurons represent entire 'chunks', concepts or templates of knowing, perceiving or cognition. These are your customer's mental structures of preconceived ideas or organized patterns of thought. Your customer's existing maps can influence and inhibit the uptake of new information. So when customers are making a buying decision, they will be processing the potential new purchase in relation to data that are currently stored within the content of their existing map framework. This process occurs at a conscious and unconscious level.

A metaphor that can help to illustrate the concept of chunking up and chunking down is that of a lake. Consider the unconscious neural maps as being under the surface of the lake. They can't be seen or observed consciously by the customer but they do exert a powerful (unconscious) influence on the customer's decision making. The customer is more aware of his or her conscious thinking that occurs above the surface of the lake where it can be more easily 'seen'. As mentioned earlier the ability to be aware of one's own thinking is known as 'meta-cognition'.

The questioning process that I will outline shortly will help customers to become more aware of their thinking processes and the maps or beliefs that underlie these. You are going to help them to make sense of and gain greater clarity over their circumstances and the challenges facing them in the achievement of their goals.

In order to chunk down we ask probing questions to dig deeper into customers' neural maps so that we can better understand their make-up and structure. To follow on from the earlier jigsaw analogy, with chunking down you may have assembled enough pieces of the jigsaw to assemble the outer frame of the problem. We are now digging down to find further pieces of the jigsaw to fill the complete picture in richer detail. We are in effect exploring these maps with the customers. Metaphorically speaking we are getting them out of their head and spreading them out on the table in front of them so that they can get a sense of perspective and understand them better. Chunking-down questions include:

- 'What specifically is it that makes XYZ company such a good supplier?'
- 'How do you know that to be the case?'
- 'How did you come to that conclusion?'

- 'What evidence do you see that lets you know that this is a problem?'
- 'What was it from your prior experience that led you to have that opinion?'
- 'What effect is this having on your staff?'
- 'What is happening that is causing a problem?'

In addition to this, chunking down can be applied to help customers understand the nature and root cause of the problems facing them. This will be explained shortly.

The other benefit of chunking-down questions is that they encourage customers to switch their focus from more external processing to more internal, inwardly focused processing. Chunking-down questions encourage them to stop and think, go inside and reflect. This can often help customers to trigger insights about the nature of and possible solutions to the challenges facing them. These are sometimes referred to as 'Aha!' moments.

From a sales perspective, an insight will often involve customers overcoming an existing assumption, or as a result of one or more of their neural maps being raised to the surface and exposed to the sunlight a better solution can present itself. On a television programme (BBC, 2013) this was described by Dr Simone Ritter from Radboud University Nijmegen as a 'schema violation', where a normal pattern of thought or behaviour is disrupted. Well-travelled neural pathways are abandoned, forcing new connections to be made between brain cells.

When customers have a moment of insight they quite literally think differently. When this creative spark occurs, a part of the brain called the anterior superior temporal gyrus activates (we have an anterior superior temporal gyrus on both sides of our brain, but interestingly it is the one on the right side of the brain that activates during moments of insight), and a burst of what are known as gamma brainwaves erupts from this area.

On a day-to-day basis we experience many different states of consciousness. The brainwave activity in each of these states of consciousness has a unique pattern that can be measured. Brainwaves change frequencies based upon the neural activity in the brain. Brainwaves are electromagnetic wave forms that are produced by the electrical and chemical activity of the brain. They can be measured with sensitive electronic equipment called an electroencephalogram. Brainwave frequencies are measured in cycles per second, or hertz, and fall into broad bands:

- *Gamma (25 to 100 hertz)*. Although the gamma range is broad, a frequency of 40 hertz is typical. Gamma waves have been discovered more recently than some of the other brainwave states. Although less is known about this state of mind, as described above research seems to show that gamma waves are associated with bursts of insight and high-level information processing.
- *Beta (13 to 30 hertz)*. Beta waves are most commonly associated with normal, wide-awake states of consciousness and a heightened state of alertness, logic and critical reasoning. Beta is increased during times of stress, enabling us to manage situations and solve problems.
- *Alpha (7 to 13 hertz)*. Alpha waves are present when people are more relaxed, when daydreaming or during light meditation. They indicate an alert state with a quiet mind. Increased alpha has been found to be present in the brainwave patterns of people who practise activities such as meditation, yoga and tai chi.
- *Theta (3 to 7 hertz)*. Theta waves are present during light sleep, including the REM dream state and during deep meditation.
- *Delta (0.1 to 3 hertz)*. Delta waves are associated with the deepest levels of physical relaxation. Delta is the slowest of the brainwave frequencies and is associated with dreamless sleep.

Neuroscientists now know that a moment before the burst of gamma brainwaves erupts from the anterior superior temporal gyrus, a burst of alpha wave activity occurs in the visual cortex at the back of the right side of the brain. This appears to momentarily 'shut down' the visual cortex, allowing the idea to bubble to the surface of conscious awareness. You may have observed that sometimes when people are thinking about a difficult problem they close their eyes to help themselves concentrate. This burst of alpha brainwave activity appears to be the brain 'blinking' to help the idea emerge into conscious awareness. So, when you ask questions that encourage customers to chunk down or go inside, allow them time to do this. Do not interrupt them by talking whilst they are thinking. Be aware that they may look down or away from you and that their eyes may glaze over or even close as they process your question. Allow this to happen.

When customers experience a moment of insight (the 'Aha!' moment), this triggers a release of dopamine and adrenalin, which creates a lift in energy and motivation. The combined effects of dopamine and adrenalin explain why, when Archimedes experienced his fabled eureka moment, he allegedly leapt from his bath and ran naked down the street!

When the moment of insight occurs you should harness this energy release and channel customers towards taking action on their insight. You may never have a better time to get their agreement to take action!

In addition, the brain rapidly associates one thing with another and particularly remembers events that are emotionally intense. An advantage is that the customer's brain will associate or anchor the insight to you and as a result of its intensity this moment will be more memorable. This helps you to position yourself as an insightful adviser to the customer!

I experienced such a moment recently when I was providing one-to-one coaching to the sales director of a large IT company. He had ambitions to become the managing director but was struggling with a relationship problem with one of his peers on the senior leadership team. As we were exploring the problem he became stuck, not knowing what to do next. He looked at me and was clearly not sure how to get out of the situation he was in. He was, if you like, circling around a well-trodden neural pathway. I decided to ask a provocative and leading question to see if I could unstick his thinking and said, 'Isn't it just time that you two sat down and talked about what isn't being said that should be being said?' He then broke eye contact with me, his eyes glazed over and he 'went inside'. I remained silent and allowed him to process. After a minute or so he suddenly sat bolt upright in his chair, his face lit up, his eyes opened wide and he pointed at me and said, 'I know exactly what to say to him! Yes, that's it! You are a genius!' We then rehearsed what he needed to say and how to say it, and the coaching session concluded. The next time we met he very excitedly told me about the conversation he had had with his colleague and how the relationship has been improved.

As I had (in his eyes anyway) achieved genius status (and sadly he is the only person I am aware of to hold that opinion of me!) he readily agreed to book further coaching sessions. I am pleased to report that he succeeded in being promoted to managing director and even more pleased to report that he has become a very lucrative client! The flash of insight was entirely his and his alone; I just played some part in helping to trigger it. However, I have benefited from the fact that it occurred and has become attributed to me.

Let us now explore a questioning framework that we can use to help us to understand the customer's context and to help customers gain valuable insight into the actions they need to take to solve their problems, overcome their challenges and achieve their goals.

The Neuro-Sell questioning map

As you learned in Chapter 4, a fundamental organizing and operating principle of your brain that drives your thinking, behaviour and action is to avoid and move away from anything that is perceived to be painful, dangerous or threatening, and to move towards anything that is pleasurable, comforting or rewarding. The 'Neuro-Sell questioning map' is orientated to and designed around this concept. It provides a structured way to question customers, to help them to become aware of the challenges and opportunities they face and to provide a dual force of 'stay away' from pain motivation and 'towards reward' motivation that will help you to close the sale.

The Neuro-Sell questioning map is illustrated in Figure 10.2, and a printable copy is available for you to download at **www.neuro-sell.com**.

FIGURE 10.2 The Neuro-Sell questioning map

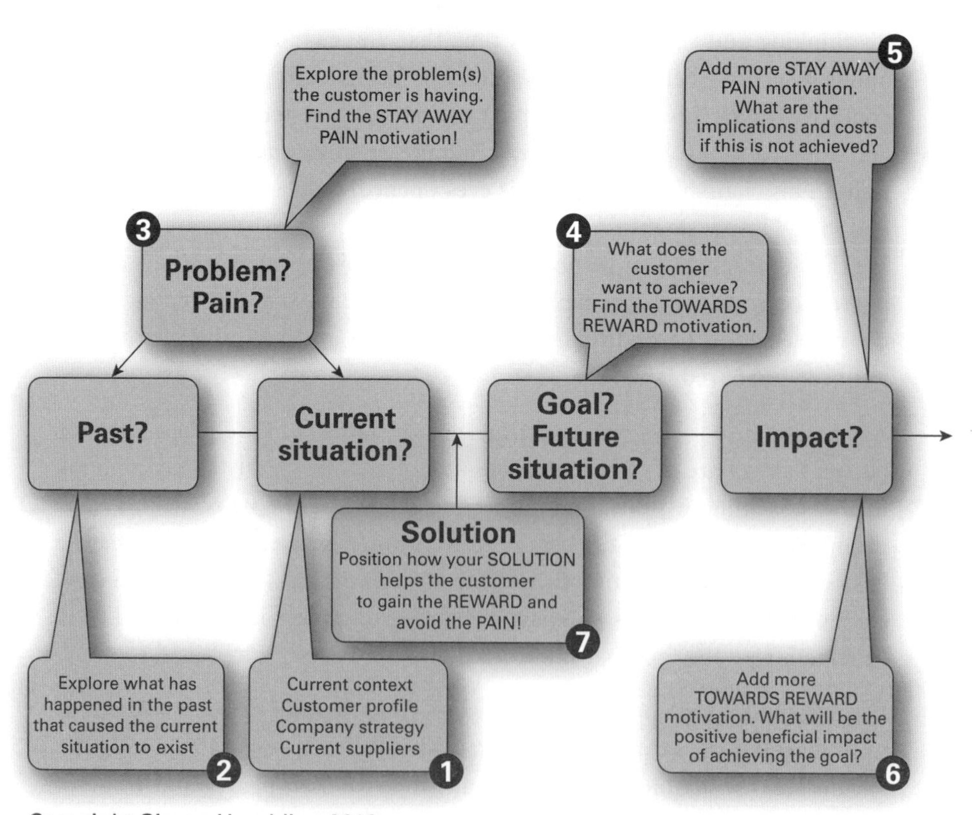

The process to follow is set out in the next sections.

1. Current situation

Where is the customer now? Question the customer to get a full understanding of the current situation. Building on the pre-meeting research that you have conducted, question further on where the customer is currently. When selling to businesses you may wish to question customers about the market they operate in, their competitors, their people, their strategy and what they are hoping to achieve in the future, current initiatives and projects, how they see their business changing and so forth. The aim of this stage is to develop a thorough understanding of their business.

If you are selling to consumers you can conduct a trimmed-down version of this to understand their personal and family situation.

2. Past

Where have they come from? It is helpful to understand customers' previous circumstances, particularly if what has happened in the past has contributed to, or is the cause of, the current situation. For example, if they are experiencing difficulties because their current IT infrastructure is struggling to cope then it would be useful to understand, for example, that part of the problem was a merger with another company that took place three years ago.

It can also be useful to understand customers' personal or career history as a way of furthering rapport with them and developing a deeper understanding of their experience (which may shape their buying approach) and to identify factors that may affect the criteria they will use to make a buying decision.

3. Problem or pain

Where are they hurting? What problems and challenges are customers facing? What isn't working as it should? What opportunities are being missed? Who or what is not performing to the required standard? What frustrations do customers have? It is important also to ask questions to begin to make customers aware of what this is costing them. This can include:

- *Financial cost.* How much is the current problem costing them? All businesses are interested in reducing cost as a way of making more profit. Where are customers wasting money? Where are they losing money? Where are they missing out on opportunities to make additional money?

- *Strategic cost.* Where is the problem impacting on customers' strategic aims? For example, if they want to expand their business how is the problem preventing them from doing that?
- *Personal and emotional cost.* What impact does the problem have on customers. Is it making them or their employees angry and frustrated? Is it wasting their time? Do customers or their employees have to work longer hours because of the problem? Does it make their life more complicated and challenging?

The aim of the questioning is to make the customer fully aware of the painful impact the problem is having. Your aim is to maximize the pain the customer is experiencing. This will provide the maximum 'stay away' motivation for the customer to move away from.

When I have been explaining this concept during speaking engagements or during consultancy assignments sometimes people tell me that this seems somewhat unethical or manipulative. My response is that what you are doing is making the customer aware of the real, genuine cost and impact of the problem. You are raising the customer's awareness of the cost of the problem(s) being faced. Sometimes people need to be faced with the brutal truth and the harsh reality of their situation before they will be motivated to take action. If the problem is causing them pain and, as a result of you making them more aware of the impact of the problem, they take an action that they otherwise wouldn't have done, removing the pain and delivering them a positive result, then I believe that you have been of good service to them. You have helped them to take a decision that proves beneficial to them. If you have to 'rub their nose' in the reality of their problem to motivate them to take positive action that improves their circumstances, then I consider that to be entirely ethical.

To help to attune your brain to identify your customer's potential problems and pain, a worthwhile exercise is to brainstorm all of the possible problems and challenges that your customers may experience. You can draw on your existing knowledge and then challenge yourself to list as many as you possibly can. As you add to this list over time it will become a valuable resource. You can use it to add to the problems the customer tells you about. You can ask if, in common with other companies in your customer's industry or situation, the customer is experiencing one of the problems. This can be positioned in this way: 'Ms Customer, a number of companies operating in your industry/area have told me that they are experiencing X problem. Is that something that affects you too?'

Firstly, this helps to build your reputation as a switched-on, knowledgeable authority within the industry in question, and secondly it may highlight further problems that add to the customer's pain, which you can sell solutions to.

It is not advisable to ever make assumptions about what a customer may be experiencing, or indeed to make any assumptions whatsoever when selling. Don't assume – ask. However, anticipating likely or potential problems can help you to be well prepared to discuss these and increase your ability to provide a solution to them with your customer.

For example, one of my areas of specialization is transforming the performance of my client's sales forces. As I have consulted with different clients I often find that many of them face similar problems with regard to their sales force performance. Here are some examples that are on my 'pain list', and I hope they inspire you to produce your own list for the industries or areas in which you sell. My 'pain list' includes:

- buyers becoming more economically cautious and procrastinating on placing orders;
- the demands from buyers for more for less;
- sales stalling and failing to move along the sales pipeline;
- profit margins under pressure and falling as a result of increased competition and more aggressive procurement practices;
- traditional, direct cold-calling failing, or becoming less and less effective;
- getting face-to-face time with key decision makers;
- poor levels of customer penetration, particularly at a senior management level;
- salespeople failing to identify the customer's decision-making unit;
- finding new customers;
- securing appointments with new customers;
- the shift from a transactional to a consultative or solution selling style;
- too many salespeople performing below expectations;
- struggling to recruit and retain effective salespeople;
- salespeople struggling to balance effort and focus;
- poor proposal to close ratios;

- struggling to retain existing customers as a result of aggressive competition;
- sales managers not spending sufficient time (or having the capability) to coach their salespeople;
- an inability to differentiate themselves against the competition;
- increasing cost of sales.

Ouch! I hope you can feel the pain! My pain list allows me to anticipate the likely problems and pain that my customers may be experiencing. If they are then I have prepared in advance to be able to discuss these and offer some possible solutions. It also allows me to ask questions around common areas of pain to see if the customer I am talking to is experiencing them too. Frequently, asking such questions uncovers additional problems my customer has that I can help with.

For all of the above I have ways to emphasize and multiply the pain (which will be covered in detail a little further ahead in this chapter), and I have planned in advance to discuss tried, tested and proven solutions to help customers with these problems. Each of my customers has a unique set of problems and challenges, and as a result I don't offer or provide a one-size-fits-all or off-the-shelf solution. However, my customers often have similar themes of problems, and anticipating these means that I am thoroughly planned and prepared to discuss these and to demonstrate how I have helped other customers to overcome every single one of these problems. I would strongly advise you to be able to do the same.

Making sure that your customers have clarity around the pain they are experiencing provides compelling 'stay away pain' motivation to their brain.

4. *Goal (future situation)*

What do customers want to achieve? Ask questions to elicit their desired outcome. Ask questions about what a successful resolution to the problem would look like and what benefits it would bring them. Be aware that they may not always be able to fully articulate this, and in helping them to become clear about what they want to achieve you will again have been of good service and value to them.

Many people in their business and personal lives can feel overwhelmed with the nature and complexity of the problems and challenges they face.

A salesperson who works with customers to get clarity about what a positive goal or outcome would be like, and to create a vision of a reality that focuses them on a new and inspiring possibility, can add great value. The more value you add, the more value customers will perceive that you and your products and services can bring, the higher price they will be prepared to pay for them.

In addition, being helpful to the customer will trigger the principle of reciprocity. Dr Robert Cialdini's research shows this to be a powerful principle of influence. In his book *Influence* (1993), Cialdini states that 'one of the most widespread and basic norms of human culture is embodied in the rule for reciprocation. The rule requires that one person try to repay, in kind, what another person has provided.'

Reciprocity is a powerful persuasion principle, and if you bring value to customers through your selling process, by helping them to get greater clarity over what would be most helpful to them, then you may very well trigger a reciprocity response in the form of them placing their business with you.

Making sure that your customers have clarity around what a beneficial goal would look like provides compelling 'towards reward' motivation to their brain.

5. Impact – 'stay away pain'

So far we have made sure we understand customers' situation, their history and what has led up to their current situation. We have explored the problems they are experiencing and the pain these are causing and we have determined a goal or future situation where the problem has been solved.

We have started to develop the dual forces ('stay away pain' and 'towards reward') that motivate a customer's brain to overcome inertia and the status quo to take action. However, we are now going to ramp up the impact of the dual motivation. To add more 'stay away pain' motivation, you now need to ask clients a series of questions to make them consider the possible negative impact on their business or personal circumstances if they do not take action to solve the problems they are facing. We need to make them reflect upon and consider the impact and costs associated with this. In step 3 we began to stir the financial, strategic, personal and emotional pain. In step 5 we must increase customers' perception of the pain they will experience if they do not take action.

Sometimes one of the strongest competitors you will face is customers deciding to do nothing, where they decide to pause and procrastinate. They will do this only if the pain that they perceive is not strong enough or the reward they will receive is not compelling enough. So make sure this is not going to happen! Ask questions to help your clients understand the full implications of not taking action. What will be the costs, particularly the financial costs, of not taking action? What will happen if the current circumstances remain? It can be helpful to get them to consider the short-, medium- and longer-term consequences of the problem to motivate them to take action. It is important to monetize the pain so that you can provide great clarity to customers about what the problem is costing them. They need to see this and feel this.

Recently I was consulting with a company selling unified communications solutions to small and medium-sized businesses. The solution they provide basically means that, whenever a customer calls the company's telephone numbers (either fixed or cellular), the system automatically routes the call to someone who can answer it. This provides a good level of customer service – it is preferable to the customer having to leave a message. If customers don't get their call answered they may phone a competitor instead, leading to loss of business.

Some of the target customers for this service were service businesses like plumbers and electricians. As part of my consultancy work, I spent some time in the field with the company's sales force meeting typical customers so that I could understand the problems and challenges they faced. It became clear that the majority of new business enquiries for such businesses came from people who had obtained the plumber or electrician's telephone number from a local newspaper advert, had seen the telephone number on the side of the plumber or electrician's van or had been given it from a friend or colleague. The potential new customer then phoned the telephone number, and the plumber or electrician took the details, visited the potential customer's home or office and then provided a quotation. The more telephone calls received the more quotations issued, and the more quotations issued the more work obtained.

I interviewed several plumbers and electricians about the challenges and problems they faced, and not being able to respond swiftly to calls proved to be a regular theme. They told me that if they failed to respond swiftly they often lost the chance to quote for the work, which in turn led to a loss of paid jobs. Using this information I developed a method of making the plumbers and electricians aware of the impact of this problem. Using the interview data I developed the following example using conservative figures:

- 10 missed calls per week;
- average call to quotation ratio: 10 calls lead to five quotations;
- average quotation to paid jobs ratio: five quotations to two paid jobs;
- average value per paid job: £750;
- £750 × 2 jobs per week × 52 weeks per year = £78,000 in lost business.

Ouch! £78,000 would be a significant loss of revenue to most small or medium-sized enterprises. You will be able to create your own examples using a similar structure.

You can also maximize the strategic and personal or emotional pain. What are the short-, medium- and long-term consequences of customers not being able to deliver on their intended strategy? What are the personal or emotional implications in the short, medium and longer term? How sustainable is it for customers to expect their employees to put up with the frustrations they are experiencing as a result of the out-of-date IT system? If the employees decide to find a job somewhere else as a result of the frustrations they feel, what will the cost be of the disruption? How long will it take the customer to replace the employee in question? How much will it cost the customer to find and train the employee?

Stir the pain! If customers thought they had a headache before, make sure that they now realize that they actually have a migraine! Give their brain a powerful 'stay away pain' motivation to avoid.

6. Impact – 'towards reward'

You also need to add a corresponding amount of 'towards reward' for the customer's brain to want to secure. In the same way that you have extrapolated the short-, medium- and long-term pain impact, now do the same by showing customers the benefits they can expect to receive. This can of course be as simple as reversing the loss or pain impact.

To return to the lost revenue example earlier, if the small business in question was currently turning over £300,000 and secured the business rather than losing it then this would be a 26 per cent increase in revenue and a corresponding increase in profit.

As you have already asked the customer about current circumstances and what the customer is hoping to achieve, you can attach this increase in revenue to the achievement of business or personal goals, for example paying off a business loan, recruiting an additional member of staff who will in turn generate more revenue, paying off a mortgage, buying a new car, going on holiday to a luxurious destination and so on. Give the customer's brain a very powerful 'towards reward' motivation.

Demonstrating and emphasizing the reward in this way will also help further on in the sales process when you discuss the price of your products and services. In the example above the cost of purchasing the unified communications solution on a monthly basis is absolutely dwarfed by the increased revenue it could generate for the customer. This makes the buying decision a 'no-brainer' (pun intended)!

The four colours of questions

In Chapters 6 and 7 we looked at how to adapt your selling behaviour to best suit that of the customer. As part of the process of adapting and of stimulating desire within each customer's brain it is useful to adapt the style and sort of questions you ask to each colour of customer. The following questions have been tailored to be attractive and stimulating to each of the four colours of customer discussed in Chapters 6 and 7.

Green customer

The Green customer will particularly respond to questions about the future, future vision, experimentation, innovating, novelty, new ideas, creating, possibility, leading the way and personal popularity, such as:

- 'As you look into the future, what possibilities have you considered?'
- 'What new solutions would you be interested to experiment with?'
- 'In what areas or ways would you like to innovate?'
- 'Have you thought about how you could create a solution to that problem?'
- 'How do you think your customers would respond if they perceived you as the thought leader in your industry?'

Blue customer

The Blue customer will particularly respond to questions about feelings, stability, certainty, teamwork, people, relationships and communication, such as:

- 'Which advantages do you think your people would like the most?'
- 'What are your feelings about the best way to ease this product into your range in a way that is manageable?'
- 'So what do we need to do to maximize people's certainty about this?'
- 'What would you most like to do to improve teamworking and communication on the back of this solution?'
- 'What positive impacts would you like to see with regard to your relationships with...?'

Red customer

The Red customer will particularly respond to questions about the form, task, results, goals, independence and speed, such as:

- 'Of the goals you mention, which is the one you want to achieve first?'
- 'Can you outline the steps you want to take to get this done?'
- 'How quickly do you want a solution to this road block?'
- 'Is it important to you that you are in control of this?'
- 'What results do you want to achieve and by when?'

Gold customer

The Gold customer will particularly respond to questions about facts, logic, rational thinking, quality, standards and detail, such as:

- 'How specifically do you know this is a problem?'
- 'What facts have you gathered so far?'
- 'What impact is this having on standards and quality?'
- 'When you analyse the data what do they tell you?'
- 'If we analyse where you are versus where you need to be what gap exists?'

Adapting your questioning in this way means that your questions will be received with a higher degree of comfort and interest, as they will be the sort of questions that the particular customer will prefer to be asked!

7. Solution positioning

Your customers will take action if their brain believes that the reward they will obtain exceeds the 'pain' that paying the money for it will cost them, or the pain they will experience from not taking action. In the example above, the reward far exceeds the cost, making the buying decision comfortable and easy for the customer's brain.

Step 7 is where you ask summarizing questions, recapping the six previous steps, to check your understanding is the same as the customer's understanding:

- where the customer is now – the current situation;
- any past influences that are having an impact on the customer;
- the problems the customer is experiencing;
- the goals or positive outcomes the customer hopes to achieve;
- emphasizing the impact of not resolving the problem;
- emphasizing the benefits of resolving the problem.

In addition to the above, before you can confidently progress any further with the sale you will also need to ask questions to make sure that you have clarity about the customer's decision-making process and who will be involved. Who is the person with the final authority to make the decision to go ahead? You will also need to know if the customer has the money or budget available to spend with you. And who is the person with the authority to spend the money?

This process is sometimes referred to as qualification. In order to qualify that the opportunity is valid and worth you investing any further time on it you need to understand: 1) cash – whether the customer has the budget available to make the purchase; and 2) authority – whether you have contact with or access to the key people involved in making the decision. Are you talking to the organ grinder or the monkey? I prefer a bold approach of asking direct questions such as: 'Mr Customer, do you have a budget set aside to solve this problem?' or 'Ms Customer, do you have a budget set aside for this?' If the answer is 'yes' then you can ask, 'Can you share that with me?'

Sometimes customers will reply that they cannot or will not share their budget with you. If this happens say 'Yes, I understand that' and then make a further attempt to flush out the budget by asking 'Can you let me know some round numbers?', 'Can you give me a ballpark figure?' or 'Can you give me some guidance as to the range within which we ought to address this solution?' You can then build on this further by saying 'Mr Customer, we have solutions to the sort of problems we have been discussing that range from £5,000 to £25,000 [whatever price range you work within]. The reason I ask is that we can solve most problems like this, but some cost £5,000 to solve and some cost £25,000. I understand your desire for confidentiality, but should I address this as a £5,000 or £25,000 solution?' This will usually result in the customer disclosing a figure of some sort, or at the very least providing some 'ballpark' amount guidance.

If the reply is still 'no' then you can respond: 'Well, that's not unusual. How do you plan to move this forward?' This bold approach is designed to flush out if the customer is really serious about moving things forward or is just looking for some free ideas and help! If this is the case then you may have to take a decision about how much more time and effort you are prepared to commit. You may be better off investing your time in a more suitably qualified opportunity.

Customers may also deliberately respond with a figure that is on the low side. This is usually done as a negotiation ploy – they hope that by 'anchoring' the opening price on the low side they will get a cheaper overall price. We will look at the whole subject of 'neuro-negotiating' in Chapter 15, but for now a counter to respond with would be: 'Hm. That could be a bit of a problem. I don't think we are going to be able to go first class on this one. I would need to take some things out of our usual solution to be able to get a bit closer to your budget.' This communicates that you have confidence in your products or services and that the only way you will be able to meet the customer's budget is by tailoring your product or service in some way to meet this.

Research conducted with people buying cars who stated that the price was 'too expensive' paid a higher price overall when they were first shown a total price and were then asked to take away optional items to reduce costs, as against people who were shown the base cost of the car and then asked to select options that they wanted to add, increasing the price as they added each option. Presenting the attractive 'deluxe' version first gives something desirable for the 'towards reward' instinct in the brain to want and to move towards. In addition, because of the hard-wired 'stay away from pain' instinct, the primitive brain has a hard-wired fear of loss. This is triggered by you describing trimming down the solution to fit the customer's budget.

By now your questioning process will have established a very powerful combined 'stay away from' and a 'towards reward' momentum in the customer's brain. In fact you will notice that the 'Neuro-Sell questioning map' switches the customer's brain back and forth between 'stay away pain' motivation and 'towards reward' motivation several times:

- step 3: 'stay away pain' motivation;
- step 4: 'towards reward' motivation;
- step 5: 'stay away pain' motivation;
- step 6: 'towards reward' motivation;
- step 7: 'stay away pain' motivation, 'towards reward' motivation, 'stay away pain' motivation and 'towards reward' motivation.

This is deliberately done to get the customer's brain oscillating between these powerful motivating factors and to create a powerful force of inner propulsion that will move the customer's brain to take positive action.

Be aware that the 'stay away from pain' threat response draws resources of oxygen and glucose away from the prefrontal cortex, which will make it difficult for the customer to consider new concepts and options. So ensure that before moving on to your pitch you saturate the customer's brain with 'towards reward' messages – the results (money, emotion, extra time) the customer will receive.

That is why you will see that the process ends with 'towards reward'. We want motivated customers in a positive, curious and receptive mood so that when we finally move to selling them our solution their brain is in the best state to receive it. We can do this using a concept called 'priming'.

Priming

Research shows that words that you have heard recently or things you have seen recently are remembered and unconsciously influence your brain and therefore your thinking and behaviour (Bargh, Chen and Burrows, 1996; Dijksterhuis *et al*, 1998; Macrae and Johnston, 1998). So, before you present your product or service, prime the customer's brain by talking about the positive aspects that are most important to the customer and the results wanted. For example, 'So you are looking for a reliable and experienced supplier, with a track record of delivering results, that you can form a long-term and productive partnership with.' The customer's brain has now been positively primed and is therefore most receptive to your sales presentation by, in this example, the use of priming words such as 'reliable', 'experienced', 'track record of delivering results' and 'long-term and productive partnership'.

So far in the sales process you haven't mentioned or discussed your solution. At this moment you have told customers what they can expect but not how. They should now be curious about how this can be accomplished. When the brain is curious it is highly receptive. It wants to know what the answer is. You can now allude to your solution but not offer it yet, by saying 'So, Ms Customer, I believe I have a solution to the problems we have discussed and one that will deliver the results you are looking for.'

Curiosity is one of the most powerful states that you can induce into a customer's brain. You will be aware that it is used to hook an audience in movies and television programmes by devices such as 'cliff hangers' where the audience is left curious about what will happen (sometimes from one show to the next to make sure you tune in next week) or maintaining a sense of curiosity about which character in the movie is the real murderer. These curiosity devices are used because they are incredibly powerful! Therefore make sure you invoke a strong sense of curiosity throughout your sales process (there will be more on this concept in Chapter 11) – your customer's brain can't resist it!

Stage 6: check

This stage of the sales process is a quick mental checklist that you need to do prior to transitioning to stage 7: convince. The checklist is designed to make sure you have covered all you need to cover and have gathered the information you need before moving into your sales pitch. The mnemonic CCAPP will help you to remember the checklist:

- *Cash.* Do you understand the customer's budget?
- *Criteria.* Do you understand the customer's buying criteria and have you prioritized them?
- *Authority.* Do you understand the customer's decision-making process and the people involved in making the decision?
- *Pain.* Do you understand the negative impacts on the customer of the current situation and the future impact if the problem is not rectified? Does the customer agree with this?

- *Pleasure.* Do you understand the positive benefits that the customer will enjoy when the problem is resolved? Does the customer agree with this?

If you can answer 'yes' to each of the questions within the CCAPP mnemonic then we are ready to move to the next stage in the 'Neuro-Sell' brain-friendly selling process and the next chapter.

11 The 'Neuro-Sell' brain-friendly selling process – the fourth phase

Convince

Perhaps the most exciting stage of the sales process for most sales professionals is when they get to pitch or present their products or services to the customer. This chapter will cover stage 7 of the 'Neuro-Sell' brain-friendly selling process and provide you with a brain-friendly pitch process that will give you an unfair advantage over your competitors.

Stage 7: convince

This stage is all about convincing the customer's brain that the action you will be proposing is a positive one, and that it will move the customer away from the pain of the problem and towards the reward that your solution will provide.

Before we go any further let us just recap on some key points about our customer's brain. We have seen that at least 95 per cent of the thoughts and feelings that influence people's behaviour and decision making occurs below conscious awareness in the unconscious mind: 'consciousness is a small part of what the brain does, and it's a slave to everything that works beneath it' (Professor Joseph LeDoux, neuroscientist, in Lehrer, 2009). Therefore it is a mistake to believe that customers will make their decision by deliberately and consciously considering the features and benefits of what you have to offer and then processing this in a logical way to arrive at a decision to buy or not to buy.

Many people make the mistake of thinking that human beings are rational, thinking decision makers and that we make deliberate and conscious decisions. The misconception is that, when we make decisions, we consciously analyse the data available, compare and contrast alternatives and carefully consider the pros and cons before coming to a conclusion.

Customers' emotions will be interwoven with their rational reasoning process. The brain has separate but connected areas for processing emotion and logic, and the combination of and communication between these areas of working influence customers' decisions to buy or not to buy. Emotion and reason are intertwined elements of our decision-making process. They influence and are influenced by each other. Indeed research shows that making good decisions will frequently involve a complex interplay of several different areas of the brain working together. It's not an emotional or rational decision. It is a blur of both – emotional impulses and more rational consideration affecting our judgement about what action to take.

We need to construct our sales message in such a way that it appeals to and is received positively by both the customer's conscious rational mind and the customer's unconscious emotional mind (see Figure 11.1).

It is also important to remember that sensory information will travel through the older, primitive brain and then the limbic or emotional brain before it reaches the rational cortex. An initial filtering process of all incoming data takes place. From a primitive survival point of view this makes perfect sense. In moments of emergency the limbic system commandeers the rest of the brain and rapidly triggers a freeze, fight or flight response. There isn't time to analyse and think; we need to act to stay alive. And the older reptilian and emotional parts of the brain are going to be the first part of the brain to receive the stimulus the customer receives from us and our sales message before it gets passed up to the cortex, where more 'rational' consideration can take place.

FIGURE 11.1 Your sales message is received by the conscious and the unconscious mind

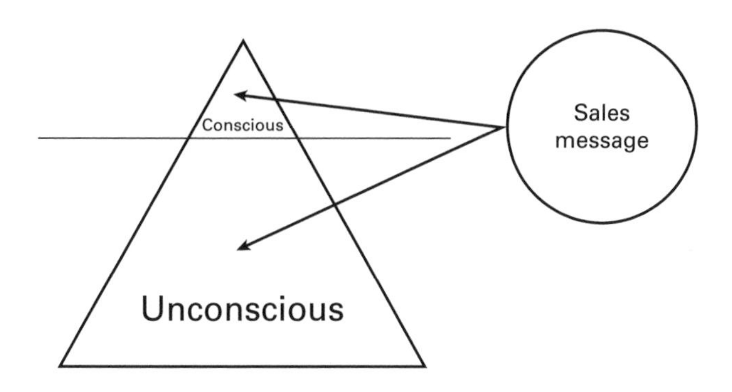

> Anatomically the emotional system can act independently of the neocortex. Some emotional reactions and emotional memories can be formed without any conscious, cognitive participation at all.
>
> (Professor Joseph LeDoux, neuroscientist, in Goleman, 1989)

The prefrontal cortex is the key part of the brain involved in conscious decision making. However, the rest of the brain is bigger and stronger! Without the other unconscious parts of the brain 'agreeing' to the decision as well, the customer is not going to say 'yes'! So we shall look at how we can structure and deliver our sales message in a way that is appealing to all three parts of our customer's brain! Let us look at the three areas of the brain and what each is interested in and will respond to.

The reptilian and the emotional brain

In Chapter 3 we looked at how the more primitive parts of the brain – the reptilian and limbic systems – are largely concerned with survival and perpetuating the species.

PRISM Brain Mapping refers to these two parts of the brain as the 'gremlin brain'. If you have seen the film *Gremlins* you will recall that when the reptilian version of the gremlins made an appearance they were nasty, destructive monsters!

The reptilian and limbic systems are the brain's chief gatekeeper or guardian, and they screen and filter what type of information will be allowed through. They pay attention to and allow through: 1) information valuable to have right now; and 2) information that alerts you to threat or danger.

These parts of the brain have little or no patience if the subject does not immediately concern well-being and survival.

They prioritize survival first (stay away from pain and danger) and then achieving comfort, so will respond to pain avoidance first. Please remember that these are largely mechanical, selfish and unconscious parts of our brain. They do not 'think' in the commonly understood definition of thinking. They rely on feelings and impressions, which can be either constructive or destructive. Their world is very black or white, bad or good, stay away or move towards reward. They respond well to clear and solid contrasts, such as before/after, risky/safe and faster/slower. Contrast helps the brain to make quick risk-free decisions.

It is important to remember that the reptilian and emotional brains judge things very quickly and without any mercy. They will form an impression very rapidly and sometimes using very little evidence. They have a tendency to focus on the negative, as their primary focus is survival, and therefore constant vigilance for threats needs to be maintained.

They like and demand instant gratification, lack self-control and restraint and will act on impulse and emotion. These parts of the brain can overreact to situations and fuel them with high and intense emotions. The *PRISM* metaphor is a useful one for us to keep in mind – we don't want to do anything to stimulate the more primitive parts of the customer's brain into perceiving us as any sort of threat or danger. If we do our sales message (and ourselves also!) will end up being rejected.

As these areas of the brain will first pay attention to staying away from danger and pain (and will put more focus on this than on moving towards reward) it is important to focus on the customer's problem first. This will capture the attention of the reptilian and limbic areas of the brain. The reptilian and limbic parts of the brain are highly attuned to noticing any changes in the environment, so consciously and deliberately moving during sales pitches, or doing something unexpected, will grab their attention.

The rational brain

The cortex and neocortex are the newest (in evolutionary terms) parts of the brain, and are sometimes referred to as the 'thinking brain' or 'intellectual brain'.

PRISM refers to the frontal lobes of this part of the brain as the 'executive brain' and the rear lobes (the parietal, occipital and temporal lobes) as the 'database brain'.

The rational brain is where working memory (or short-term memory as it is also described), problem solving and more logical, analytically based decision making take place. Its world is less black and white than the reptilian or emotional brain, and it is involved in considering possibilities and analysing data, discovery and logical thought. It will attempt to ascertain the facts and the truth. When it has achieved this it will attempt to put things together in a logical and structured manner and determine what action needs to be taken.

This part of the brain also has a conscience, and will consider ethical principles of what is right and fair, making judgements about what is the correct course of action to take.

Unlike the impulsive older parts of the brain, this part can choose to delay rewards or gratification. The rational brain seems to be fulfilled by achievement and satisfaction, which generally speaking are dependent upon having a sense of purpose that involves having a direction and meaning to one's life.

The database brain has two main functions – to think and act automatically using programmed thoughts and behaviours, and to provide a reference source for stored data, values and beliefs. The various parts of your brain have put data into this area of your brain during your life. This part of the brain does not have any original thinking or interpretation power, but acts on the information it has stored. The neocortex, emotional brain and reptilian brain access the database brain to see what previous data (in the form of experiences, memories and beliefs) are available. These data will then influence what action is taken.

How useful the data are depends on how accurate or valid they are. You may at times find yourself selling to a brain that has retrieved previously stored data that are not helpful to your sales attempt. For example, customers may have a belief about the sort of product or service you are selling based on inaccurate data. They may have heard someone say negative things about your sort of product or service. Your job will be to understand and then influence the stored data that are being used to underpin their belief.

The good news is that research shows that memories are very plastic and malleable and can be altered, changed and adapted (Sanitioso, Kunda and Fong, 1990). In Chapter 10, we looked at probing and chunking down into the unconscious thinking of your customer and surfacing more of it to help the customer make a better decision. We will build on this through the rest of this chapter. It is always useful to understand the beliefs and opinions your customers have now (based on the data stored in their database brain) so that we can meet them where they are before taking them somewhere else.

What follows is a series of considerations to use and incorporate into the structure and content of your sales pitch.

Curiosity

The very first thing you must do with your sales pitch is to capture the attention of your customer's brain. And then you have to keep its attention.

If the brain encounters anything new, novel or unusual it pays attention! When people experience anything that is different or unexpected, norepinephrine and dopamine levels in the brain rise. This causes people to focus their attention and makes them alert and interested.

The best way to capture the attention of the customer's brain is to surprise it. Some of the methods that I have used when selling various forms of consultancy, sales training, negotiation training and management or leadership development include:

- Decorating the meeting room the client was using to hold the sales pitch with materials we use when running sales training programmes. I re-created the stimulating, engaging learning environment I create for programme participants with props like wall posters containing key learning points, inspiring quotations, music playing, and cards on the desk and floor that participants use during exercises on the real programme. I then invited the potential client to join us in the meeting room to start the pitch. The people who were sitting on the client's decision-making panel were very curious about how unusual the meeting room now looked! I didn't make any reference to anything that I had adorned the room with until much later in the pitch, thereby maintaining the sense of curiosity and the client's attention.
- Having an acronym or mnemonic that contains the first letter of six key concepts I will discuss displayed on a flip chart, and as the pitch progresses I complete the words that match the first letter. I make sure I leave at least one incomplete until really close to the end of the pitch, as the client's brain is crying out to close the gap!

- Starting with a provocative and/or challenging question or statement, for example 'In your company how poor a level of performance can you operate at and still keep your job?' or 'According to research at least 80 to 90 per cent of employees' behaviour is determined by the behaviour of the company's leaders. This means that the behaviour I see your employees exhibiting tells me what sort of leaders you are.'
- Using a prop or device. As much of my sales consultancy is orientated around the application of neuroscience, one of my regular travelling companions is a scale replica model of the human brain. I will often have this sitting on the table or desk in front of me and deliberately not make any reference to it until much later in the pitch. As it is an unusual object it attracts the attention and curiosity of the client.
- Walking in with a giant full-colour graphical storyboard that provides a visual representation of a solution I can provide. The entire pitch was conducted using this and this alone. The client's logo and references to their customers and employees were contained on the storyboard, which showed that it had been prepared specifically for the client. I left the storyboard with the client at the end of the pitch at their request.
- Telling them early in the pitch that I will reveal several pieces of information that are compelling and interesting to them. I open what I call a loop in their brain that they want to close. I don't close the loop until much later in the pitch, as a way of building curiosity and keeping their attention. For example, 'As we progress I am going to show you three tried and tested and proven ways that we can deliver exactly the results you are looking for and provide you with rock-solid evidence of our ability.' As I say 'three tried and tested and proven ways' I count off three times on my fingers. As I move through the pitch and reveal each 'tried and tested and proven way' I mark each point verbally ('So the first way is...') and non-verbally by marking off each point on the relevant finger. This reminds the client's brain non-verbally that there is yet more to be revealed. I will explain later in this chapter why I choose 'three tried and tested and proven ways' and why I specifically structure the phrase as 'tried and tested and proven ways' rather than 'tried, tested and proven ways'. So you will have to wait a short while to find out the specific and powerful reason for that. Hopefully I now have your attention and curiosity!

So please give some thought to how you can capture your customers' attention, and make their brains curious to know more. Trigger the release of norepinephrine and dopamine to focus their attention and make them alert and interested. You don't have to be wild and wacky to surprise the customer's brain. You can do this in subtle ways that will still be very effective. Mysteries, puzzles, questions, unexplained things and incomplete patterns are all powerful ways to grab and keep the attention of your customer's brain. Indeed, please be aware that being too unusual and different can trigger a sense of anxiety in the older parts of the brain, as too much novelty could mean a big change, and change can be perceived as a threat to survival.

As mentioned earlier the reptilian and limbic areas of your customer's brain are highly attuned to notice any changes in their environment, so moving during sales pitches or doing something unexpected will grab their attention. You can use movement to make sure you don't lose the customer's attention. I would recommend doing something to attract attention every few minutes or so. Examples include, if you are using visual aids such as PowerPoint or Keynote, blanking the screen and walking across to the other side of the room, alternating between standing up and sitting down, passing something to the client to look at, moving closer to the audience to make a key point, asking a question, illustrating a point on a flip chart or whiteboard, displaying a visually impactful graphic or photograph before providing the reason for showing it, and so forth. When practising your pitch, plan to insert attention grabbers throughout it.

Chunk

To provide the customer's brain with comfort and certainty chunk your sales pitch into bite-size chunks or simple steps. Outline the process that you are going to follow with the customer at the start of your pitch. This will establish a mental pathway, which makes your pitch easier for the customer's brain to process. This primes their brain to expect each step (this helps with comfort and certainty) and allows you to stimulate curiosity (as outlined above) by mentioning things that you will be telling the customer about – only not just yet!

Clarity

It is important to provide the customer's brain with clarity. To help customers make a decision we need to provide clarity throughout the sales process and particularly during the pitch. The greater the clarity we can provide the easier it is for their brains to draw conclusions. As we will see later, when we provide the brain with clarity rather than clutter and confusion it is far easier for it to process the information it is presented with and make a good decision.

Conscious thinking is a complex interaction that includes billions of neurons, and as a result the brain uses a lot of energy when doing it. As a result of the brain evolving at a time when food could be a scarce commodity, it has evolved to be as efficient as possible and to minimize energy usage where possible. When making decisions and solving problems, the brain makes heavy use of the prefrontal cortex. The prefrontal cortex can become overwhelmed when faced with confusing decisions to make. When faced with such a scenario the brain may avoid making the energy-sapping decision or make an automatic or unconscious choice. This is quicker and easier and uses less energy. The danger is that the automatic or unconscious choice could be just to keep things as they are, to stick with the existing supplier or to procrastinate on making the decision. If you can minimize the energy usage by the prefrontal cortex you will maximize the energy resources that the customer's brain has available to make a good decision that will serve well.

Salespeople who understand their product or service very well can sometimes, because of their higher level of insight and years of experience, explain things at a more advanced and abstract level than suits a person with less knowledge and experience. When this happens customers can become confused and feel somewhat bamboozled. Their brains are finding things too complicated and confusing and as a result decision paralysis can result.

There is a limit to how much information the rational or conscious brain can process simultaneously. A variety of researchers have concluded that the maximum number of 'chunks' or 'bits' of information a person can keep in mind simultaneously is between four and seven, with one study concluding that the number of 'chunks' of information that you can recall accurately at one time is – one (Cowan, 2001; Gobet and Clarkson, 2004)! The fewer variable factors that the customer's brain has to hold in mind to be able to make a decision needs to be limited as much as possible. The 'Neuro-Sell' mantra is: less is more!

The customer's capacity to make a wide decision is limited by the resources available to the prefrontal cortex. The aim of the 'Neuro-Sell' approach is to help the customer by making the buying process as brain-friendly as possible. We can achieve this by providing clutter-free clarity to the customer!

A good way to provide clarity is to start with where the customer is now. In Chapter 4 we looked at mental frameworks or maps that exist in our customers' brains. It is a good idea to recognize where customers are now with their thinking about the topic in question, what they believe and the opinions they have.

It is important to make sure that we are 'on the same page' as our customers or, more accurately, sharing the same map! Showing that we understand their situation and beliefs helps us to build rapport with them. It is good to summarize your understanding and check it is correct with customers. Getting them into an agreeable frame of mind by agreeing with your explanation of their current context primes their brain to being more receptive to agreeing with your sales proposal.

We can then use this current situation or context to demonstrate how we can move customers from where they are now towards a more beneficial situation. The brain works by connecting new incoming information into existing maps, so linking your ideas or proposal to the customers' existing reality will make it easier for them to understand and accept.

When people face a problem or challenge they usually attempt to apply strategies that have worked for them in similar situations previously. Customers will be projecting their experience from the past on to the current or future problem or challenge as a way of dealing with it.

Their belief about what might work could be a barrier to your proposal being properly considered and accepted. It is usually more effective to meet them where they are currently in their thinking (showing that you understand their current 'map') and then build upon this and gracefully move them towards a new positive solution than it is to tackle their existing 'mental map' head on.

The first step in making sure customers have clarity is to be certain you have clarity yourself, that is, that you are able to articulate your proposal in a short, sharp, brain-friendly manner, communicating the most important points with impact. A useful framework is to use a technique I was told about by a communications consultant who trained company executives how to make a positive impression when being interviewed by the media. He told me that it was important that they were very clear about the main messages they wanted to get across, and I have found his framework to be very valuable for sales professionals to use to consider the key messages in selling situations – both in formal pitches and on more informal occasions.

You can adapt this to use earlier in the sales process as part of your pitch, and you can also use this to summarize the key benefits or to provide your suggested options.

He told me that he uses the metaphor of a house – what he called 'the message house' (see Figure 11.2). The 'roof' of the message house contains your most important point. This is a point that you make reference to several times, for example at the beginning, middle and end of your presentation. When building your sales presentation you need to give careful consideration to this message. This message is going to be the number one thing that you want your customer to remember from your presentation. Make your main message orientated around something that provides a strong benefit to the customer. For example, I often use the fact that I have a practical results-orientated approach, agree success measures in advance of doing any work for a customer and guarantee to deliver the agreed results. In a nutshell my main message is: 'You want improved sales results – I guarantee to deliver them!' As simple as this may sound, this frequently provides a point of contrast to my competitors, who appear to spend more time talking about the service they provide rather than the results the customer desires.

Your main message is then supported by your three key points. These can be, for example, the main three benefits that you can provide. These should

FIGURE 11.2 The message house

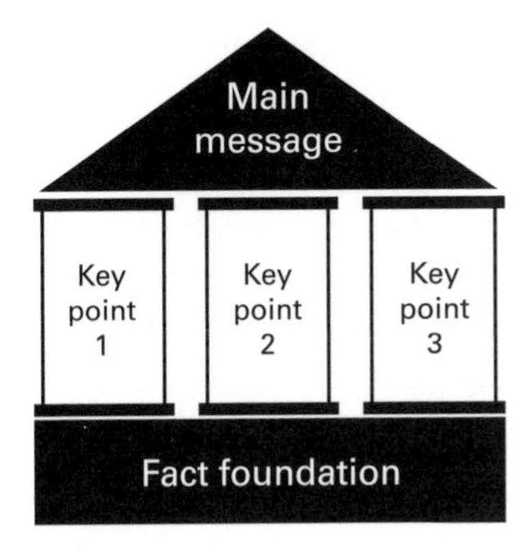

link up to and meet the customer's three most important criteria that you carefully identified during the context and catalyse stage of the 'Neuro-Sell' brain-friendly selling process (see Chapter 10). Owing to the limitations about the number of items or chunks of information that can be held consciously at any one time, one main message and three key points are a good number to provide. This number can be processed, understood and decided upon by the customer's brain.

A little earlier I mentioned that I would also explain the reason for using the structure of 'tried *and* tested *and* proven' rather than 'tried, tested and proven'. The reason is that using the two *and*s adds rhythm, power and emphasis to your message by separating and punctuating the three key points. Political speech writers know this technique, and now that you are aware of it you will see it being used quite often!

You then need to prepare your 'fact foundation'. These are facts and proof points that you have ready to prove the validity of your main message and three key points. Although we know that emotion plays a significant role in buying decisions, we still need to have rock-solid proof of the capability of our products and services.

Please remember that certain types of customers will need more facts and data as part of their decision-making process (the Gold personality preference, for example), and there is also a strong emotional benefit to providing this proof, as it provides a sense of certainty, reliability and comfort that the brain wants to move towards.

Contrast

Your customer's brain (including the all-important older parts) responds positively to clear contrast. To help it make the right decision provide it with contrast, including contrast between what the customer's situation is now and what the situation will be like once the customer has chosen to purchase your product or solution. Your contrast needs to provide:

- A summary of the customer's current state or situation, the problems being experienced and the costs or impact associated with these. This provides the 'stay away pain' motivation for the customer's brain to want to move away from.
- A summary of the desired future state situation that shows the rewards and benefits that the customer will experience once the problems have been solved by the purchase and implementation of

FIGURE 11.3 Your product or service enables the customer to move away from the problem towards a positive solution

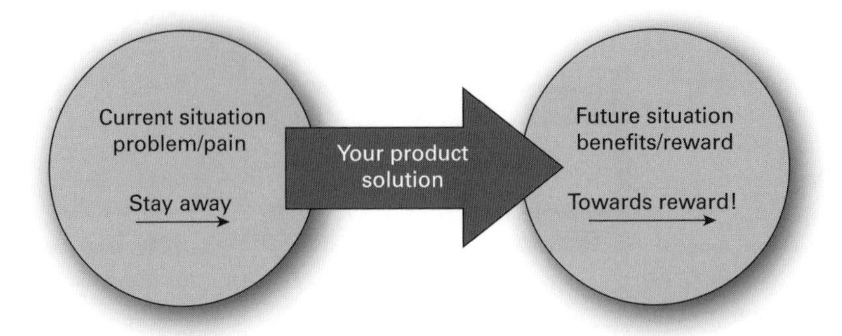

your product or service. This provides the 'towards reward' motivation for the customer's brain to want to move towards.

- A positioning that shows your product or service as the enabler that allows movement from where the customer is now to where the customer wants to be (see Figure 11.3).

My good friend and key account management expert Phil Jesson has devised a powerful framework that you can use when working with and selling to customers. It is called 'strategic bridges'. The framework shows the current state of the customer's business and the desired state of the customer's business within a defined timescale, for example within three years. Metrics or measurements of where the customer is now and where the customer wants to be are included, for example turnover, margin, etc.

You then determine the strategies that the customer is planning to use to take the business from where it is now to where the customer wants it to be. These are the 'bridges' that will enable the customer's transition away from the current state and towards the desired state (see Figure 11.4).

In order to be able to develop and articulate this framework, excellent, in-depth customer knowledge is required. Although it may take time to develop, the rewards are an exceptional insight into the customer's business, which will set you apart from your competition.

In some instances you may have to work with the customer to develop the strategic bridges framework. You can add huge value by helping the customer construct it, and this deepens and strengthens your relationship.

FIGURE 11.4 Strategic bridges example

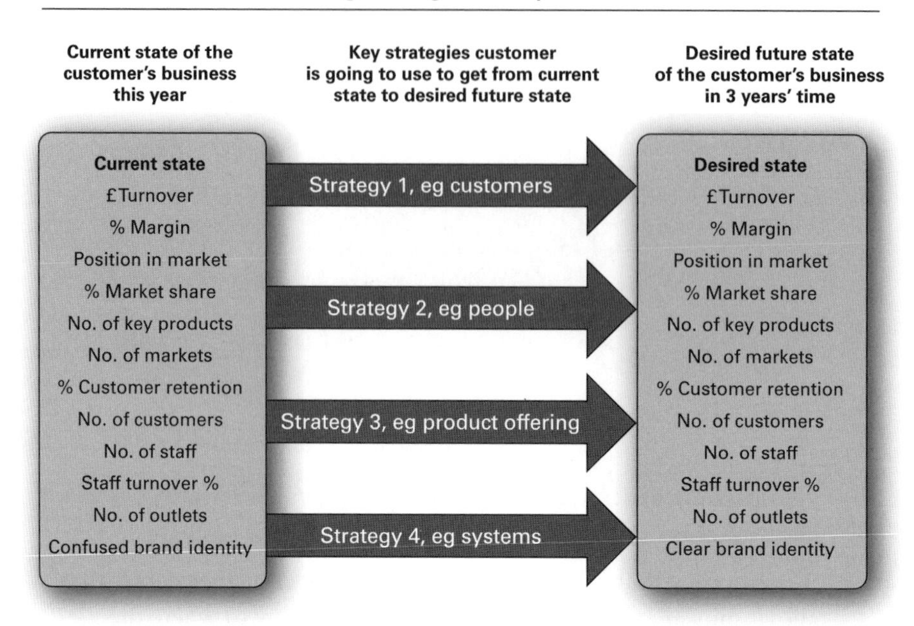

I have used the strategic bridges framework on numerous occasions when selling into my own customers, and the response has always been very positive. It shows that you understand – really understand – their business and that you can help them to achieve the results they desire.

You can then use the framework to discuss and agree how your products and services help customers to successfully implement their strategies. There will usually be a series of projects and initiatives aligned to each strategic bridge, and if you can add value to these a sale will usually be the result!

You move yourself from being just another supplier to being regarded as a strategic partner and strategic enabler, with all of the rewards that will bring to you. You differentiate yourself and put clear blue water between yourself and your competitors.

You will also notice that the strategic bridges framework builds on the 'stay away from pain' and 'towards reward' principle discussed earlier. You can blend the pain and problems of the current state and the results and rewards of the desired state into the strategic bridges framework to create a powerful and sophisticated brain-friendly selling tool. Based on personal experience I cannot recommend it highly enough.

A further point of contrast is to demonstrate how you differ from and are superior to your competitors. A good way to find out is to ask your existing customers why they buy from you and the advantages they perceive you to have. What you may think differentiates your organization might be different to what your customers think! If you ask your key customers your points of difference then you will rapidly develop clarity about what they are, as common themes will occur. You can then incorporate these into your sales pitch. Your message needs to clearly differentiate you from your competitors and provide a strong point of contrast.

For example, one of my customers contracted me for a piece of consultancy work. They are a FTSE 100 business with operations across the world and were about to reorganize one of their largest divisions. As part of this they wanted to merge three separate sales teams into one new global sales operation. They wanted me to help them to do this and to ensure that their new global salespeople had the required skill set. I was delighted to be told that this was not a competitive situation, as they were talking only to me about this project. They explained that this was because 'You have a really in-depth understanding of our business and you have a proven track record of delivering results.'

This is very nice feedback to receive, but most importantly it is powerful customer feedback on how they perceive my points of difference that I can then use with other customers, differentiating myself by focusing on my ability to consistently deliver tangible results and performance improvement. While the majority of my competitors talk about how great their training programmes and consultancy are, I focus on what my customer most wants – improved results.

Later in this chapter we will look at how we can use this feedback from customers in a very powerful and compelling way that will help you to close more sales – but more of that later.

Concrete

If I can't picture it, I can't understand it.

(ALBERT EINSTEIN)

One of the challenges that many organizations face is the transition that their salespeople must make from transactional selling to consultative or solution selling. Owing to the competitive nature of most industries, margins are gradually eroded as companies fight each other using price as the key point of differentiation. If customers want to purchase a product and perceive no clear point of differentiation between two suppliers, they will invariably place their business with whoever has the cheapest price. And the margin spirals invariably downwards.

In order to differentiate themselves and protect their margin position companies increasingly have to shift their focus to be providers of tailored solutions rather than just transactional suppliers of products. This has necessitated a shift from the more traditional 'box-shifting' salesperson to the more consultative solution-orientated salesperson.

One of the inherent challenges that accompany this is the salesperson's ability to sell less tangible and complex solutions, and finding or training salespeople to sell successfully in this manner is a major point of pain for many sales directors.

Your sales pitch needs to be as concrete as it possibly can be. Intangible and abstract ideas are difficult for the customer's brain to process. Such thinking is usually the domain of experts who are capable of processing in this manner.

As described earlier, when making decisions and solving problems the brain places heavy demands on the prefrontal cortex. The prefrontal cortex can become tired and overwhelmed when faced with confusing decisions to make. So make your pitch as concrete as possible. Concrete language and concrete explanations are required for most customers' brains to grasp and understand a concept. If you are selling a tangible product bring your product (or at least part of it) with you. Let the customer touch and feel it. If you are selling something more intangible make it more concrete by providing a visual representation of it or what it does, for example a flow diagram. Showing customers something is far easier for their brain than them trying to imagine it for themselves.

About 25 per cent of the brain is involved in visual processing (more than any other sense). Make sure you capture the brain's attention with strong visuals. A message accompanied by a picture is far more memorable. In order to consider and weigh up a complex decision, the visual circuitry of the brain is frequently activated, so providing a more concrete visual input can help the customer to make an effective decision.

Visual images (such as flow charts and diagrams) can contain a lot of information and are very brain-friendly, as they reduce the amount of information the brain needs to use to take in the information. The use of visual imagery can reduce the demand on the prefrontal cortex, which leaves it better able to process information and make a decision.

Certainty and credibility

Your customer's brain likes certainty. At a deep level the primitive part of the brain links certainty to survival. A hundred thousand years ago the more certain your environment was, the safer it was. Uncertainty felt like a threat to survival. Fast forward 100,000 years to the present day and your customer's brain is still constantly automatically and unconsciously seeking certainty in preference to experiencing uncertainty. Our brains are always looking to move away from the discomfort of uncertainty towards the comfort of certainty.

Human beings are creatures of habit. We follow the same routines, travelling the same way to work each day, sleeping on the same side of the bed, eating the same sort of food and watching the same sort of television programmes day in and day out. Yes, we also like some variety in our lives, but we have a deeper-seated need for certainty.

Our affinity with branded goods is linked to a certain degree to the certainty that these branded goods will provide. We feel comfortable that they will deliver what we are expecting, which increases our sense of certainty, leading to the higher psychological levels of comfort that our brain craves so much.

Your customers will want to be as certain as possible that you are a safe and reliable supplier who will deliver on your promises. Customers may be feeling somewhat insecure, nervous, exposed, concerned or that they are taking a risk when choosing a supplier to work with. They are feeling uncertain. You must communicate a strong degree of certainty.

Firstly, it is vitally important that you come across as an experienced, knowledgeable professional who knows what he or she is talking about. You must become an expert in your field with exceptional levels of knowledge in your products and services. You must ooze authority and confidence (but no overconfidence) from every pore of your being. Customers have to buy you first before they will buy from you.

Secondly, your sales pitch must contain certainty and credibility drivers. Certainty drivers are things that you can use to drive customers' levels of certainty upwards. Credibility drivers show customers that you have done it before and are capable of helping them. Certainty and credibility are intertwined. The more confident customers are in your credibility the more certain they will feel, and the more likely they are to buy. Examples include:

- *Testimonials from existing customers.* Dr Robert Cialdini's (1993) research references the power of what he calls 'social proof'. People use the behaviour or beliefs of other people to decide how to behave and act themselves. The more uncertain people feel, the more likely they are to use the behaviour of others for guidance. In addition, Cialdini found that people are more inclined to follow the lead of people who are most similar to them. So you may need to develop a series of powerful testimonials that allow you to match them successfully with the customer to whom you are selling. Large companies need to see testimonials from large companies. Small companies need to see testimonials from small companies. If you can match industry then that is even better. I recently closed a piece of business to provide sales manager coaching and sales training for a large international company. The training will be delivered in several languages across a number of countries. Therefore the testimonial I used was from an existing customer where we had worked successfully on that scale of project. Perhaps the most powerful form of testimonial I can recommend is for you to connect the new customer with one of your existing customers directly. Allow them to talk about you and to meet together without you being present. The power of this form of testimonial has closed many deals for me. Sometimes just the fact that you have the confidence to offer to connect them directly with one of your existing customers is enough for the potential customer to feel a sense of certainty. Even though it is obvious that you are only going to connect the potential customer with a happy and loyal existing customer who will say positive things about you, the transparent nature of the offer to connect them to discuss directly is powerful.
- *Client lists.* In addition to specific testimonials prepare a comprehensive list of the sort of organizations you work with. I know from analysing data from the website I use to promote myself as a keynote speaker that, after looking at what topics I speak on, people usually visit my client and testimonial page. They are looking for certainty.

- *Case studies.* These are basically more in-depth versions of testimonials. They need to be short enough to be read by a busy potential client but have enough depth to prove your capability and track record. Be aware that when you design them you should provide a big-picture summary for the Green and Red customers, and provide a finer level of detail and proof for the Blue and Gold customers. Your case studies need to explain the client's situation, the solution you provided and the results you obtained. Include specific people-orientated examples for the Blue Customer. Your client will need to be quoted and featured throughout the case study to build the social proof.
- *Research from a recognized authority that proves the efficacy of your product or service.* This can be a very powerful certainty driver. Be prepared to provide in-depth data to your Gold customers, who will take the time to analyse the research and check the sources are accurate.
- *Guarantees.* These are another way to provide certainty. The all-too-familiar money-back guarantee is a strong certainty driver. If you don't deliver the client doesn't pay. This also communicates certainty, as unless you were certain of your capability you wouldn't provide the guarantee in the first place. The marketer Jay Abraham popularized the concept of the 'risk reversal', where the business takes all the risk away from the customer by providing a rock-solid, 100 per cent no-quibble money-back guarantee. This has been proven time and time again to grow sales.
- *Endorsements from people who are perceived to be influential or authority figures.* Celebrity endorsements are commonplace, and that is because they work. I have secured endorsements from famous, high-profile businesspeople for my books. For example, multimillionaire Duncan Bannatyne, star of BBC TV's programme *Dragons' Den*, wrote the foreword to my book *Bare Knuckle Negotiating* (Hazeldine, 2011a), and multibillionaire entrepreneur Michael Dell endorsed the book *Bare Knuckle Customer Service* (2012), which I wrote with my co-author Chris Norton. This is good for book sales, as the endorsements from these authority figures provide certainty that the books are worth reading.
- *A pilot programme, study or test.* This allows clients to 'dip their toe in the water' first before committing fully. It allows you to prove your capability and minimizes the risk to the client – very appealing to Blue and Gold customers!

Make sure you have an armoury of certainty drivers in your sales arsenal. Your customer's brain needs certainty – your job is to provide it.

The power of 'story-selling'

Another way you can add certainty is through telling stories. I call this 'story-selling'. This is such a powerful method of selling in a wider sense that I am going to devote a separate section to this.

The method builds upon the research of Robert Cialdini about the power of social proof and then expands and leverages this concept by harnessing one of the most powerful and compelling forms of communication – the story.

Stories are found in countless aspects of human life, including speeches, books, writing, education, songs, films, television, video games, theatre and art. Stories play an important part in human culture. They are a ubiquitous component of human communication. A study by anthropologist and evolutionary biologist Robin Dunbar (1998) found that social topics, for example social activities, personal relationships and personal likes and dislikes, which he broadly referred to as 'gossip', accounted for 65 per cent of speaking time among people in public places, that is, people largely telling stories about themselves and other people!

Human history is rich with stories. Before the written word, stories were how learning and wisdom were passed from generation to generation. Many religious books and religious teachers make extensive use of stories to get their teachings across. Human history is also rich with legend, fable and metaphor. Stories were probably one of the earliest forms of entertainment. They have a lasting impact. Aesop's fables, for example, can be traced back at least to the 4th century BC – proof of their longevity and power.

Most people's early experience of story is a positive one. From bedtime stories, listening to grandparents talking about the good old days, and reading fiction, to watching stories on television and in the movies, our upbringing has been saturated with story.

Now neuroscience allows us to understand exactly why stories are so powerful and, most importantly for us as sales professionals, how we can use them to inform, educate and influence our customers' brains.

We all seem to enjoy a good story, but what is it about stories that makes them so engaging? Why, when we listen to a narrative of events, for example, do we seem to get pulled into the story?

In experiments conducted by Dr Uri Hasson and his team of neuroscientists at Princeton University a woman tells a story whilst being monitored by an fMRI scanner. A group of volunteers then listened to the story through headphones while they had their brains scanned by the fMRI scanner. What is fascinating is that, when there was activity in one part of the woman's brain, corresponding activity occurred in the listeners' brains. When the woman had activity in the emotional brain region the listeners did too. When her prefrontal cortex activated, so did theirs! When she was telling her story she was planting ideas, thoughts and emotions into the listeners' brains. When you listen to stories and understand them, your brain 'dovetails' with the brain of the storyteller. You experience the same brain activity as the storyteller.

Hasson also recorded a graduate student record an unrehearsed story about a disastrous high school prom that involved rival boyfriends, a fight and a car crash, again whilst undergoing an fMRI scan. A group of 12 people then listened to the recording while undergoing fMRI. Again the results show that the speaker and the listeners had similar brain activity.

So, when you are telling your customers stories, the neural activity between your brains synchronizes, increasing understanding, comfort, rapport and connection. When you tell a story to customers, you can transfer experiences and emotion directly to their brains. They feel what you feel. They empathize. Anything that you have experienced (or that one of your other customers experienced and is being transmitted via your story), you can get your customer to experience the same. This phenomenon is likely to be related to the mirror neurons described in Chapter 3.

In addition, whenever we hear a story, our brain attempts to relate it to one of our existing experiences. Listeners connect the story with their own ideas, memories and experiences. Stories engage our customers' brains, drawing them into the narrative, inviting them to join in and play along. A good story will elicit similar emotional and motivational responses to the real experience being described. In the same way that, when we read a fiction book, we are drawn into the author's world and re-create the experience of what is being described inside our brains, your customers will be drawn into your selling story.

With stories you will not have a passive, uninvolved audience. Your customers cannot just listen to your story – they will be drawn in because your story deliberately activates their brains. Research shows that when people imagine seeing flashing lights this activates the visual area of their brain, when people imagine someone tapping on their hand the tactile areas of their brains are activated and when people imagine looking at the Eiffel Tower in Paris their eyes instinctively move upwards as though they were actually looking up at the real Eiffel Tower! And if you take a short moment to imagine that you have lemon juice in your mouth and imagine swilling the sharp and sour taste around your mouth you will notice that you are now producing increased quantities of saliva!

Stories also work in print, as anyone who has read a good novel will testify. Stories about your existing customers, if well crafted, will stimulate the reader's brain. The brain does not make much of a distinction between reading about an experience and experiencing it in reality. In each case, the same regions of the brain are stimulated.

Structuring your selling story

Your customers' brains are most likely to be stimulated by your story when customers are drawn into the story and transported along by it. The events need to unfold one after the other, with a clear demonstration of cause and effect. This matches how our brains like to think.

What follows is a story-crafting framework that you can use to develop your very own range of powerful selling stories. In keeping with the brain's hard-wired tendency to stay away from pain or discomfort and to move towards reward your story needs to incorporate these motivating forces. The structure is a narrative that presents a series of connected events with a clear cause-and-effect format. This has been captured in a structure called the 'Neuro-Sell story-selling map' (see Figure 11.5). A downloadable copy is available for you free of charge at **www.neuro-sell.com**. Here are the elements we need to have in place for a successful sales story:

- *Customer.* If at all possible select a customer whose circumstances are as similar as possible to the new customer to whom you are selling. Get as close to the personal circumstances, company size and industry as possible. For this reason you may need to develop a bank of stories to suit different customers to whom you are selling. To minimize any push-back about the lack of similarity you can inoculate against this by saying: 'I appreciate this customer isn't exactly like you, as every one of our customers is unique. However, I do think this customer is close enough for you to get an idea of what we do.'

FIGURE 11.5 The Neuro-Sell story-selling map

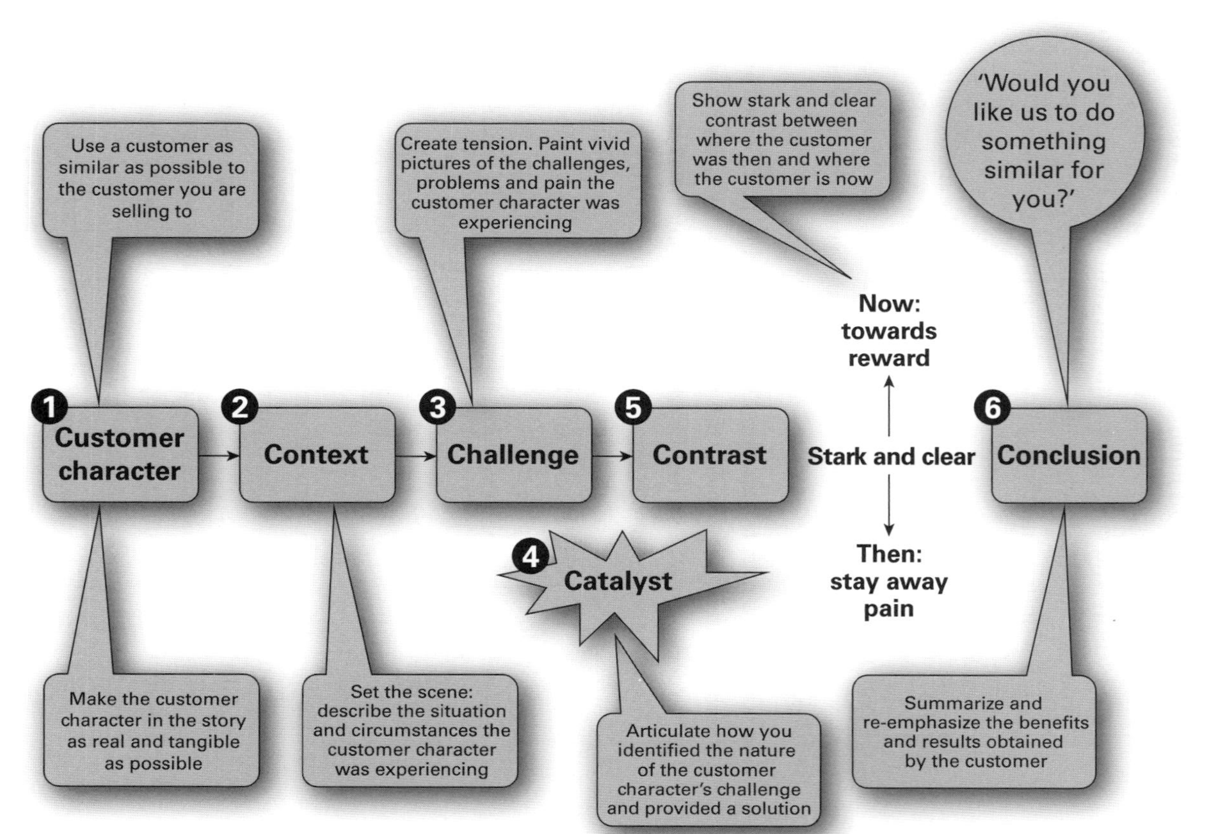

- *Character and characterization.* Make the customer character in the story as real and tangible as possible. If you have permission to do so, then use his or her actual name. If not, then say something like 'For reasons of customer confidentiality I can't mention his real name so let's just call him John.' The more you can tell about the customer in the story, the more the new customer you are selling to will be able to identify and empathize with that customer. Throughout your story you can add to the character. Characters may be presented by means of description and through describing their actions, speech or thoughts.
- *Context.* Set the scene for your story. Describe the situation and circumstances the character was experiencing. Make this as vivid as you can. Draw the customer who is listening to the story into the world of the character, so that their (and your) brains begin to synchronize.
- *Challenge (create tension).* You now need to paint a picture of the challenges, problem and pain the character customer was experiencing. Describe the issues the customer was facing; label and describe the negative emotions being experienced – pain, anguish, stress, despair, hopelessness, frustration. A very common plot in books, TV programmes and films is the 'challenge plot' where the underdog triumphs, the hero's rags-to-riches story is played out or someone triumphs over adversity. These types of stories where characters overcome obstacles and reach their goals are very appealing, and this sort of plot is very powerful when creating selling stories. If you want to add great power and impact to your story, as you describe the emotions the character customer is experiencing, experience them yourself. As you describe the feeling, experience some of it yourself. Demonstrate it in your voice, on your face and with your gestures and other non-verbal communication. The mirror neurons in the listener's brain will be activated by this. This is providing a strong 'stay away' motivation that will trigger the customer to take action.
- *Catalyst.* This is where you and your products and services make an appearance. It is your intervention that precipitates the positive change that the character customer has been looking for. Articulate how you identified the nature of the customer character's challenge and provided a solution.

- *Contrast.* Vividly describe the journey the character customer went on from a place of pain and problems to a positive and pleasurable solution. Draw a very stark and clear contrast that articulates the benefits your product, service or solution has brought to the customer. Describe the gains made in money, emotion and time. Again, as you describe the emotions the character customer is experiencing, experience them yourself. This is providing a strong 'towards reward' motivation that, combined with the existing 'stay away' motivation you stirred up earlier, will trigger the customer to take action. Emphasize the results you achieved for the character customer in the story.
- *Conclusion.* Conclude with a very brief summary of where the customer was and where the customer is now. Re-emphasize the results you helped the character customer to achieve. You can then ask if the customer you are telling the story to would like you to do the same for him or her.

You need to keep your story short and sweet. Keep it to between one and three minutes at a maximum. Any more than this and the customer's attention may start to wander! Here is an example:

> Mr Customer, perhaps the quickest and best way for me to let you know what we do is to tell you a short story about a customer we helped a few months ago. Ms X is the managing director of an XYZ business that turns over £X million. I appreciate they aren't in exactly the same business as you, but I do think they are close enough for you to get an idea of what it is that we do to help companies like yours improve their sales performance.
>
> Ms X contacted us because, as in so many other businesses, her profit margins were being squeezed by aggressive competition and her customers driving a hard bargain. Despite the fact that her sales team are quite experienced, they just didn't seem to be closing as many deals as they used to. Even her most experienced salespeople were struggling. In short, their pitch to win ratio was falling off a cliff. And because of her salespeople feeling increasingly desperate they were starting to give in to customers' demands to lower their price. So they were winning less business and making less profit. Ms X was tearing her hair out. She had tried everything she could think of to improve the situation, but nothing seemed to work. She was getting very nervous that she would have to downsize her company in order to survive.
>
> When we met with her we applied our unique sales force effectiveness assessment framework to her sales process and her salespeople. It became apparent that the weak link in the chain was her salespeople's ability to differentiate her company from the competition and to be able to negotiate more margin into the deals they closed.

> We then agreed some robust commercial outcomes with Ms X so that we all had clarity on what success would look like. We held a series of interactive workshops with her sales team and, using our unique neuroscience-based pitch structure, helped them to produce value propositions that really differentiated them from their competitors. We then took them through some demanding negotiation workshops using highly realistic simulations to build their confidence and competence. And to make sure the learning was applied and embedded we scheduled weekly virtual review sessions with them using our online training platform to follow up on their plans and to provide ongoing coaching.
>
> Within three months we had helped Ms X's sales team put £4.6 million of revenue into their sales pipeline and added 3 per cent to their bottom-line margin. Ms X can now sleep soundly at night knowing that her business is in growth and that her dreams of expanding the business can be realized.
>
> So that's it in a nutshell! We helped a struggling business transform its sales capability and as a result improve revenue and profit, moving it from a frustrating and difficult situation to a business whose salespeople walk confidently into sales pitches and close deals with good margins.
>
> Now I do appreciate that your situation is unique and that the ideal solution for your sales force will therefore look somewhat different, but I would love to discuss with you how we could work with you to improve your sales performance and your results.

I hope that this example inspires you to create your own series of sales stories!

Stories such as this appeal to, and impact upon, the conscious and unconscious parts of your customer's brain. If you have a well-constructed plot for your story the customer's conscious brain will be following the plot line and be curious about how the person in question overcame the challenges faced. The unconscious part of the brain is responding to the emotional experiences that the story stimulated and is drawn to the journey in the story that moves the person in it from a state of discomfort to one of comfort.

Stories help people to understand and remember. They make a lasting impression. I was called recently by the sales director of a company who wanted me to deliver my 'Bare Knuckle Negotiating' keynote speech at his company's annual sales conference. He told me that he had seen me speak at an industry event a month ago and 'just loved my stories'. My keynote speeches are deliberately crafted to contain a series of real-life stories that I use to illustrate key points and to educate and entertain my audience. It is interesting to note that it is my stories that he most remembered, and as a result he wanted to book me to deliver a speech to his sales force. Indeed I have met people who have seen me speak several years ago and they will often tell me that they most remember 'the story about when you...'. Be memorable with your customers and customers-to-be – tell them stories.

If all has gone according to plan, customers should now be convinced that you are the correct supplier for them and that your products or services will provide them with the solution they are looking for. Your final task it to get their commitment, and we will look at this in Chapter 12.

12

The 'Neuro-Sell' brain-friendly selling process – the fifth phase

Close the deal

It has been said that if you can't close sales then fundamentally you can't sell. This chapter covers the eighth and final stage of the 'Neuro-Sell' brain-friendly selling process and is devoted to the all-important matter of getting the customer to make a firm commitment and to place the business with you.

Stage 8: confirm and conclude

As we move towards the end of the sales process we are nearing the time when we are going to ask the customer to make a decision.

Please remember that fundamentally people buy: 1) a solution to problems ('stay away from pain'); and 2) good feelings ('towards reward'). Therefore it is important when you begin to conclude your pitch to summarize: 1) 'stay away from pain' motivation – remind them of the pain they want to move away from by making the right decision; and 2) 'towards reward' pleasure motivation – remind them of the pleasure and comfort they can move towards by making the right decision.

Invoking the fear-of-loss principle described in Chapter 10 will also provide a powerful stimulus to take action. Research by Daniel Kahneman and Amos Tversky at Hebrew University showed that the pain of loss was approximately two times more powerful as a motivating force than the pleasure of gain or reward and that these forces were powerful determinants of people's decisions. They named the concept 'loss aversion'.

> In human decision making, losses loom larger than gains.
>
> (Kahneman, Slovic and Tversky, 1982)

So make sure that first you emphasize the loss that will be incurred if nothing is done to improve the current situation and remove the problems and the effects they are having on the customer. Stir in a healthy dose of longer-term loss impact – 'Based on the figures we have discussed I calculate that over the next two years you would miss out on about £175,000 in lost revenue' – and the 'stay away from pain' pot should be bubbling nicely!

Now provide a reminder of the attractive nature of the future scenario that customers will experience when they have made the right decision, the problem and attendant pain have been removed and they are experiencing the outcomes, pleasure and comfort of your solution. You may wish to make use of pleasure and reward images in your sales presentation. When the brain is expecting a reward (such as food, sex, money or positive social interactions) it generates dopamine in anticipation – hence the profusion of reward imagery being used in advertising! I am not suggesting that you litter your sales pitch with pictures of scantily clad models, but good-quality photos showing pleasurable outcomes and results (money, achievement, smiling or accepting faces, awards, status, etc) will be attractive to the reptilian and limbic brain.

Dopamine is the drug of desire and is important to moving the brain into a 'towards reward' state. To maximize the transition from 'away' to 'towards', maximize customers' sense of certainty by minimizing threat (by emphasizing certainty as described earlier) and then show the rewards they can expect. Making reference to rewards will help to trigger dopamine release and desire for your product. Research has shown that an unexpected or surprising reward releases more dopamine than an expected reward, so it is a good idea to keep a benefit in reserve and add this when customers aren't expecting it. For example, when you are summarizing your offer, wait until you have finished, check with the customers if that meets their needs and then add the extra unexpected benefit by saying, for example, 'Oh, and I forgot to mention that when you place your order you will also get...' This will trigger a larger release of dopamine and attendant desire, which will be helpful to you when you ask for customers' final commitment.

An important tipping point is that people will also buy when they are convinced that the benefits they will gain exceed the cost of purchase. That is why your sales pitch has to strongly communicate value and a quantifiable return on investment. It has to show the customer's brain that the benefits will far outweigh the 'pain' of the purchase price!

Provide the customer's brain with a clutter-free, clear next step and use this to confirm actions on both sides and to conclude or close the deal. In my experience, the actual final concluding or closing of the deal usually comes down to nothing more complicated than asking customers if they are comfortable with going ahead, and agreeing what needs to be done in order to make this happen. The 'close' does not have to be (and shouldn't be) some dramatic crescendo that occurs at the end of the sales process. On the contrary, I advocate checking in with customers throughout the sales process, getting feedback from them on how they are feeling about what is being discussed. The process I follow is to:

1. ask testing questions;
2. ask trial conclusion or closing questions;
3. ask final conclusion or closing question.

Testing questions should be used throughout the sales process and test or check that you are on track. Examples include:

- 'How do you feel about that?'
- 'Does this make sense so far?'
- 'Does this seem to be going in the right direction?'

Trial conclusion or closing questions are used to test customers' comfort with making a decision and taking action. You are not asking them for their business just yet; you are checking out that when you do ask they are going to say 'yes'! Examples include:

- ‘Is this what you were looking for?’
- ‘If we can get the specification you require, how close are we to going ahead?’
- ‘If we could do that would you be interested?’

Trial conclusion or closing questions will also help to highlight any areas of concern or hesitation that customers may have.

The final conclusion or closing question is used to get the customer’s final commitment, close the business and take the order. As a result of the testing and trial conclusion or closing questions you have asked so far you will have a good idea of the customer’s comfort level.

I monitor customers’ comfort and interest levels throughout the sale, paying close attention to their verbal and non-verbal behaviour (more on this in Chapter 14), looking and listening for positive indicators that they are interested, feeling comfortable and ready to make a decision. In the sales literature these are referred to as ‘buying signals’ and include nodding in agreement, positive language or sounds such as ‘Yes’, ‘Uh-huh’ or ‘OK’, leaning towards you and your product or presentation, brightening up, talking faster and more excitedly about your product or service, reaching out to touch the product, making notes, calculating numbers and asking questions.

Once you have moved through asking testing and trial conclusion or closing questions, and you are seeing and hearing a number of the positive indicators described above, then this would be a good time to ask a final conclusion or closing question. Examples include:

- ‘Would you like to go ahead now?’
- ‘Would you like to try it?’
- ‘Shall we get started straight away?’
- ‘When do you want to get started?’

The three-step questioning process combined with a careful monitoring of customers’ comfort and interest levels is very effective. It does rely on you asking customers for their agreement. The fundamental principle is that once you are certain customers are interested and comfortable you should then move to conclude or close the sale. Be bold and ask for the business. Never leave a customer meeting without asking for the business or, in the case of longer or more complicated sales cycles, concluding or closing on a clear and positive action step that moves the sale forward. You either leave with the order or you leave having demonstrably moved the sale forward. That is what sales professionals do.

Limit choice if you want them to make a decision

If customers' brains become overwhelmed with too many possible choices and variety available to them then this can cause them to hesitate and to feel less certain, with the result that they are likely to procrastinate. Having to choose from a profusion of choices is confusing. Too much choice causes confusion, and confusion causes inaction. Give them something to buy!

It is important to strike the right balance between providing a degree of positive choice (the brain likes this, as it gives it something to move forwards to and gives it control over what it chooses, and therefore a feeling of certainty) and providing a profusion of choice that will be counter-productive.

Research by Sheena Iyengar and Mark Lepper (2000) from Columbia and Stanford universities respectively found that 'people who had more choices were often less willing to decide to buy anything at all, and their subsequent satisfaction was lower when they had been confronted with 24 or 30 options than when they faced six options'. Iyengar and Lepper add that 'choice, to the extent that it required greater decision-making among options, can become burdensome and ultimately counterproductive'.

Neuroscientists know that there is a limit to the amount of information that can be held and processed simultaneously in the brain. As mentioned previously, a number of different pieces of research have concluded that the number of items or 'chunks' of information that your customers are able to keep in their minds simultaneously is limited. Therefore the fewer chunks of information, and the fewer variable factors that customers have to hold in mind when making a decision, the easier it will be for them to make that decision.

Provide a limited choice of three main options as a maximum. This is brain-friendly, as it gives customers a sense of autonomy and control and makes the buying decision easier. If you overwhelm them with choice they may choose the 'do nothing' option.

Once customers have made the main option decision, it is then easier for their brains to tinker with or make alterations to the characteristics of the main option than to choose from a helpfully intended but overwhelming sea of options. The sea of options is intended to provide maximum choice and therefore possibility of purchase. What it does instead is to overwhelm the prefrontal cortex of customers, drive their levels of uncertainty up and lead to decision paralysis or procrastination.

If you have followed the process successfully so far, you will have secured the customer's business and closed the deal. However, just to make extra sure, Chapter 13 contains even more powerful brain-friendly selling tips for you to use to gain the advantage over your competitors.

13
Some more brain-friendly selling tips

Just to make doubly sure you get the business, here are some further areas to consider when designing and delivering your sales pitch.

Being memorable

In some situations you may have to present or pitch to a customer (or a committee involved in making the buying decision) and not be able to get a decision on the day. Customers may be seeing several suppliers or wish to consult with everyone involved in making the decision before going ahead. In these cases you need to make sure that you, your presentation and your products and services are memorable. The brain particularly remembers: 1) primacy (the first thing it sees or that is said), so you need to start your pitch or presentation with a strong opening 'hook' or benefit that will be remembered; and 2) recency (the last thing it sees or that is said), so you need to finish your pitch or presentation with a strong summary of your key benefits and differentiating factors.

A common occurrence for some sales professionals is to participate in a competitive pitching situation where the final shortlist of suppliers are invited to present one after the other on the same day. The brains of the customers have to sit through a series of supplier presentations and then make a decision on which supplier to select. This puts a big demand on the energy resources of the brain.

This also has application for which part of the day is a good time for you to make your pitch. Ideally you want to be the first or the last supplier presenting. If you end up with a slot in the middle ground, you are likely to be less memorable and less impactful. You may also wish to consider that at the end of the day the customers' brains are starting to get tired and that this will be impacting on their ability to receive and process your message.

Although I have successfully won competitive pitches at all times of day, my personal preference is to be the first supplier, as the customers' brains are at their most receptive, and it provides the opportunity to shape their perceptions about what they need in a way that favours my proposal!

For example, in a recent pitch, I was selling our sales performance capability and services into a global digital advertising company with a colleague. We knew that one of our competitors was an industry specialist. The competitor knew the client's industry very well and had a track record in it. I succeeded in getting our pitch scheduled first in the day and, having anticipated (correctly as it turned out) that the competitor would make much of their industry knowledge, I boldly stated that we were not specialists in the world of digital advertising. I said that we were specialists in improving sales performance and that we apply our experience of transforming sales forces from and to a variety of industries, bringing fresh ideas, fresh perspectives and world-class best practice to our clients. I said that as part of our methodology we would want to gain a solid understanding of the business and the challenges the client faced, and that we had a robust consultative process that would help the client to properly understand what sort of sales force was needed to deliver the strategy. In doing so, I managed to prime and shape the customer's perception of what they wanted (eg the advantages of having a fresh and unbiased perspective) and when they phoned to confirm that we had won the business specifically referenced our ability to bring a fresh perspective that would differentiate their salespeople from those of their competitors!

The following help with being memorable:

- *Being unusual.* Doing or saying something unusual (within the confines of common sense and good taste!) will get you remembered.

- *Repetition.* Things that are repeated tend to stick in the memory, so articulate your key benefits and points of differentiation several times – at a minimum towards the beginning of your pitch, during it and at the end of it.
- *Emotion.* Events associated with strong emotion are more memorable, and you can communicate emotion in terms of your own enthusiasm and passion, and by going into and modelling the emotional states involved in your customer story as described earlier. The mirror neurons in your customer's brain will get fired up, and the emotion is transferred from your brain to that of the customer, making you and your pitch more memorable.

Keeping it simple

Make your pitch simple, straightforward and easy to understand. The customer's brain likes to conserve energy, so don't make it have to work too hard. Keep your explanations clear and jargon-free.

Keep any printed or visual materials clean and crisp, with plenty of 'white space'. Keep them clutter-free so as not to distract or confuse the customer's brain.

Distracting or irrelevant information is only going to exhaust prefrontal cortex resources and interfere with the rational decision-making capability. If the prefrontal cortex becomes overwhelmed, the customer's brain will resort to making automatic or unconscious choices, which may or may not be helpful to you!

Making changes

Conclusive research about what the average attention span of an adult is has been difficult to locate, but the research I have conducted indicates that it is somewhere between five and 10 minutes. So it would appear prudent to plan to make some sort of significant change or movement every five minutes or so throughout your pitch.

Owing to the brain's orientation reflex (an automatic and involuntary response to a new stimulus that results in increased brain activity and increases in adrenalin), any significant change or movement will recapture the full attention of the customer's brain. So deliberately build in several stimuli such as blanking your PowerPoint slides off if you are using them (by pressing the letter B or W when in screen display mode) and stepping in front of them to make a key point, turning your PowerPoint slides on, moving towards or away from the audience at key stages, walking from one side of the room to the other (more on this a little later), tapping on the presentation, flip chart or whiteboard with a pen, standing up and handing something to the customer at specific stages, and so on. To keep your customer's attention keep firing stimuli at them at regular intervals that will activate the orientation reflex!

Using metaphors

A metaphor is a short cut to understanding; it can make it easy for the customer's brain to understand your message.

In a metaphor, one thing is compared to another. We understand new or complex things in relation to things we already know. This can help to make complex or unfamiliar things easier to understand. For example, at a networking event I met someone who was a joint venture broker, who acted as the middle person between two companies and helped them to work together in return for a percentage of the increased business. This person used a metaphor to describe the activity involved, which was 'We are a dating agency for businesses!'

You can use metaphors in a number of ways. For example, in price negotiation situations I sometimes use metaphors such as 'You can't pay Ford prices and expect a Rolls-Royce', and in sales pitches you can emphasize product benefits using metaphors such as 'Our new cutting-edge production process guarantees that the edges of the product will be as smooth as silk!'

If your clients use metaphors at any stage of describing what an ideal solution for them would look like, for example 'It's got to run like clockwork!', then 'steal' the metaphor and make reference to it, for example 'As you mentioned, this will run like clockwork!' Clients' metaphors are meaningful to their brains, so replay them back to the clients and link them to a benefit of your proposal.

Going multi-sensory

The more senses you can deliberately incorporate into your pitch the better. The more sensory-rich your pitch is, the more engaging it will be to your customer's brain.

Seeing

Use strong, rich, vivid, clear visuals and consider where you place them. Research by Dr AK Pradeep (2010) of NeuroFocus found that:

> placing images on the left and words on the right is superior for rapid processing by the brain. This is because items in the left visual field are perceived by the right frontal lobe, whereas the right visual field is perceived by the left frontal lobe. Since the left frontal lobe is specialized in most people for interpreting semantics, while the right frontal lobe is specialized to process imagery and iconography, this speeds up processing and contributes to a positive emotional impression.

Hearing

Vary your voice pitch, tone, volume and speed to add interest. Pause for effect after making a key point. Ask powerful questions that engage the customer in the presentation.

Touching

Touch is the oldest human sense and the most urgent and internal to our survival and confidence. Wherever you can include a tactile element do so. Give the customer something to hold, examine, weigh up or explore. If handing out written material ensure it is printed on heavy, glossy, good-quality paper that feels attractive.

Smell and taste

If you are selling anything that can incorporate these additional senses then make sure that you do so. Although not directly connected with my pitch or what I am selling, I will at times make use of these senses. For example, when scheduled to pitch last at the end of four other potential supplier presentations, we arrived armed with a box of cream cakes. I positioned with the customer that we were aware that they had been listening to supplier presentations all day long, and as we were the last we didn't want them to be losing attention so the cakes were there to help blood sugar levels! This was all done in a humorous manner and had the added advantage of being somewhat novel and memorable to boot. We got the deal.

Spatial association

One aspect to how the human brain works is its tendency to swiftly associate one thing with another thing. Indeed one of the key aspects of how human memory works is by association. For example, a certain favourite piece of music may be strongly associated with a specific occasion or person. There is a piece of music that is always played when my favourite ice hockey team come on to the ice at the start of a game. The state of excitement that exists just before a game has become associated with this particular piece of music, and any time I hear it my brain starts to re-create the excited emotional state that it has linked to the music.

It is possible to associate, for example, almost any emotional state or feeling with some form of sensory 'anchor' that you link to it. For example, the visual stimuli or anchor of a blue flashing police light in your rear-view mirror will tend to be associated with a particular state of mind and feeling!

A powerful use of this property of the brain is spatial association. Part of my stagecraft as a professional speaker is to associate certain places or spaces on the stage with certain times, people, emotional states and so forth.

An example that works very well in sales pitches is to anchor or associate a current or past challenging or problematic situation with the left-hand side of the area you are presenting with – to the left of the screen where you are displaying your PowerPoint graphics, for example. You will associate this part of the room in the customer's brain with the problem being experienced, and it will be the place and situation that provide the 'stay away from pain' motivation. You then associate or anchor the right-hand side with the future solution. You will associate this part of the room in the customer's brain with the pleasure, comfort and results that will be experienced, and it provides the 'towards reward' motivation. (See Figure 13.1.)

You create the 'stay away from pain' association by discussing and describing the problems, challenges and pain customers are experiencing with them (demonstrating the negative emotional state yourself whilst doing this) only

FIGURE 13.1 Spatial association

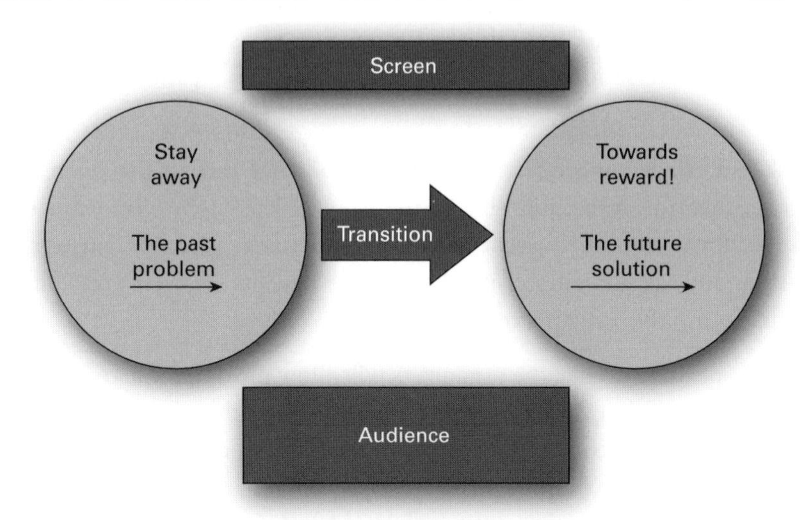

whilst standing or sitting in this location. Customers' brains rapidly come to associate that area with the negative emotional states associated with the problem. You create the 'towards reward' association by moving from the 'stay away from pain' area across to the other side of the room where you have decided to locate your 'towards reward' association location. When explaining and highlighting the benefits, using a positive, confident upbeat emotional state, you do this only from this location. Customers' brains rapidly come to associate that area with the positive emotional states associated with the solution.

You can also associate the transition between these two states by describing the process you will follow or the journey you will take with customers whilst walking from the 'stay away from pain'-associated location across the front of the room to the 'towards reward'-associated location.

Many people's brains will unconsciously code time along a mental line, with what happened in the past over to their left-hand side and the future over to their right-hand side. This appears to be the way that the brain codes and relates to time. They will gesture, for example, up and to their left when describing events from their past and gesture up and towards the right when describing events in their future. For this reason the current (soon to be past!) problems the customers are experiencing are spatially located to the left of the audience and their new bright future is spatially located to the right.

This unconsciously sends the message that you can help customers to make the transition from their current negative problem to a positive future solution. Once you have established a positive 'towards reward' location, deliver all conversations about the solution from this location in the room only.

If asked a challenging question from the audience during the pitch, step into the 'neutral' transition area in the front of the room (you want only positive 'towards reward' associations with the future location), answer the question and, as you emphasize the positive benefits of what you can do, step back into the 'towards reward' location.

At some speaking events, the organizers will allow me or ask me to make the audience aware of my books, products or services so that they can purchase them if they want to. Using the above spatial association when talking about the problems, challenges and frustrations that I or my customers have experienced in the past, I do so from the left-hand side of the stage (from the audience's point of view). I then describe the journey or transition that solved the challenges as I walk slowly across from the left of the stage (from the audience's perspective) to the right of the stage, where I conclude by discussing the new, positive outcome that was achieved.

For example, when talking about my learning journey as a negotiator from a raw, inexperienced beginner, who got regularly 'beaten up' by his customers, to the current situation where I have written a best-selling book on the subject, and speak and consult on the topic across the world, I associate the past to the left, describe the learning journey I went on and then describe the current, more confident and capable negotiator I have become from the right of the stage.

The confident, capable state is associated with the right-hand side of the stage from the audience's perspective. And it is from this exact spot that I describe and offer my books and products to the audience. The audience has seen me play out my learning journey, showing the experience that has gone into my current state of capability and confidence. Their brains swiftly associate these positive states with my books and products, which always results in a queue of people wanting to see me after my speech to make a purchase!

Please be aware that this entire process operates outside of the audience's conscious awareness. It influences them at an unconscious level. It is however a very deliberate and calculated strategy on my part. And it should be for you too – it is very powerful.

Giving the customer's brain something to complete

When I am asked to send customers a written proposal there are two things I always do. I was introduced to these two very powerful ideas by Chris Norton, with whom I wrote the book *Bare Knuckle Customer Service* (2012).

Firstly I call the proposal a plan, so rather than it being titled 'Sales training proposal' it is entitled 'Draft sales training plan'. I will usually call it a 'draft plan', as this invites the customer to become involved in it and contribute to it. This increases the customer's sense of involvement and ownership.

This bold approach communicates confidence that we will be proceeding with the 'plan', and allows me to include the second vital element – a strong action summary at the end. I do this in a very specific way. Table 13.1 is a genuine example that shows the action summary I included in a written proposal to provide a sales training programme for a new client. All I have done is to edit it to remove any specific client references for reasons of confidentiality.

If you take a look at this you will notice several things:

1. It details all of the action steps that need to be taken, who is responsible for them and when they will happen by. This helps to increase the client's sense of confidence and comfort that everything will happen as agreed.
2. It includes the all-important evaluation of delivered results, which again helps to increase the client's feelings of confidence and comfort.
3. The first three actions are ticked and shown as being completed. This gives a sense of action and progress, despite the fact that at this stage the client has not given the final go-ahead! It also leaves a list of actions that need completing. The brain likes closure and completeness. As mentioned earlier, leaving things undone or incomplete (only the first three actions have been ticked as complete) creates a desire for closure in the customer's brain.

TABLE 13.1 Action plan example

Action step	Timescales	Action owner(s)	Completed
Initial telephone call to discuss requirement	3 October	Shirley Smith Simon Hazeldine	✓
Meeting to further understand needs and requirements	7 October	Andy Jones Simon Hazeldine	✓
Draft action plan reviewed	11 October	Simon Hazeldine	✓
Draft plan reviewed	17 October	Shirley Smith Andy Jones	
Agree commercial objectives and measures	20 October	Shirley Smith Andy Jones Simon Hazeldine	
Programme content, format and timeline agreed	20 October	Shirley Smith Andy Jones Simon Hazeldine	
Familiarization of Simon Hazeldine with customer products and services	27 October	Shirley Smith Andy Jones Simon Hazeldine	
Programme design and materials produced	10 November	Simon Hazeldine	
Programme dates scheduled	10 November	Shirley Smith Andy Jones Simon Hazeldine	
Brief line managers of participants and provide programme overview for their subsequent review and follow-up	14 January	Simon Hazeldine	

TABLE 13.1 *Continued*

Action step	Timescales	Action owner(s)	Completed
Programme run for Group 1	1 to 3 February	Simon Hazeldine	
Programme evaluation review	10 February	Shirley Smith Andy Jones Simon Hazeldine	
Programme run for Group 2	7 to 9 March	Simon Hazeldine	
Programme evaluation review	16 March	Shirley Smith Andy Jones Simon Hazeldine	
Review progress, behavioural changes, sales pipeline status and coaching requirements with participants' line managers (Session 1)	20 April	Simon Hazeldine	
Review progress, behavioural changes, sales pipeline status and coaching requirements with participants' line managers (Session 2)	25 May	Simon Hazeldine	
Review meeting to measure progress against agreed commercial objectives and agree next steps	31 May	Shirley Smith Andy Jones Simon Hazeldine	

In a study by consumer researchers Joseph Nunes and Xavier Dreze (2006), loyalty cards were handed out to 300 customers of a local car wash. Every time they had their car washed their loyalty card was stamped, and when the card was full the customer got a free car wash. Two different types of cards were issued – one stated that 10 stamps were required to receive a free car wash, but two stamps had already been placed on the card. The second type of card stated that eight stamps were required to receive a free car wash, but no stamps had been placed on the card. In reality both cards offered the same incentive – if you buy eight car washes we will give you a car wash free of charge. It was only the way they were communicated that was different.

The data were fascinating. After several months, only 19 per cent of the customers who had received the blank eight-space loyalty card with eight spaces to fill had made enough visits to claim their free car wash. However, 34 per cent of the customers who had received the 10-space loyalty card with two stamps already included had made enough visits to claim their free car wash. And this group took less time to complete their eight purchases than the other group!

According to Nunes and Dreze, perceiving the loyalty programme as something that has been started but is incomplete rather than something that has yet to begin motivates people more to complete it.

Although my action plan summary is not a loyalty card I believe a similar effect is being produced. My customer's brain feels motivated to complete the plan, which is great news for me!

So there we have it: you have successfully journeyed through the Neuro-Sell brain-friendly selling process. Please take the time to apply what you have learned and incorporate it into your current selling process. Take action on what you have learned and you will reap the rewards!

In Chapter 14 we will explore the fascinating area of body language or non-verbal communication and what hidden messages the customer's brain is sending us on a constant basis about the customer's readiness to do business with us.

14 Body language and the truthful brain

You may be asking why a book about the brain and selling contains a chapter on body language or, as it should more accurately be described, non-verbal communication. Selling is fundamentally about the process of communication, and (although research data and opinion vary) somewhere between 60 and 80 per cent of communication is non-verbal.

The part of the brain that is primarily responsible for non-verbal communication is the limbic system. You will recall that the limbic system is one of the more primitive parts of the human brain. It is unconscious and reactive. For this reason it is sometimes referred to as the 'truthful brain', as the non-verbal communication it generates (at least initially) has not been filtered or influenced by the higher-order thinking of the cortex. It reacts instinctively to what happens in the world around it and generates non-verbal communication based upon this.

Spoken communication can be considered and structured in advance of delivery. Non-verbal communication is generated by the limbic system in real time and is delivered unfiltered and without any conscious interference. As a result, it has more purity and can provide us with a powerful insight into what someone is really thinking and feeling. The customer's non-verbal communication can provide us with powerful cues that we can use to adapt and alter our sales behaviour to maximize our chances of success.

So, to begin, let us define non-verbal communication. Fundamentally anything that isn't words can be classified as non-verbal communication, including:

- body movements;
- eye contact;
- facial expressions;
- gestures;
- interpersonal space – or proximics (mentioned in Chapter 9);
- paralinguistics and vocal cues – tone, inflection, pauses, intonation, volume, pace, pitch;
- posture;
- touching – of self and others.

Although some non-verbal communication is cultural (for example, the same gesture such as a 'thumbs up' means different things in different countries) and some is idiosyncratic, the indicators discussed in this chapter will be limbic in origin.

In Chapter 9 we began to look at non-verbal communication and how to use it to make the customer feel comfortable with us, so that the instinctive threat response was not triggered. In this chapter we will be going more deeply into non-verbal communication, focusing on becoming more conscious of it, so that we can gather information that is useful to us.

Non-verbal communication is a fascinating subject that could easily fill several books, so this chapter will focus on some of the key things to be aware of from a sales perspective.

A key theme throughout this book has been to focus on and utilize the innate, hard-wired 'stay away from pain' and 'towards reward' circuit in the human brain. This chapter will develop this concept further, identifying non-verbal communication that will tell us when customers are feeling comfortable ('towards reward') or uncomfortable ('stay away') with us and what they are hearing. If we pick up non-verbal indicators or displays of comfort then this gives us feedback to carry on. However, if we pick up non-verbal indicators or displays of discomfort then we need to slow down and consider what we are doing.

When you have been in a car you may have experienced the effect of rumble strips. Also known as sleeper lines or audible lines, they are a road safety feature that alerts drivers to potential danger on the road by causing a vibration and audible rumbling that is transmitted through the wheels of the car into the car body. They are often placed across the road, and when you are approaching a junction or roundabout the rumble strips warn you to slow down and take greater care. Once you are attuned to the non-verbal indicators of discomfort these will act like 'rumble strips' that tell you to slow down and make sure you do something to make the customer feel more comfortable.

In addition, it is important that you make sure that your non-verbal communication displays high degrees of comfort and confidence. If you look comfortable then this sends a powerful and largely unconscious message that you are confident about your products and services and are comfortable that they will do what you say they will do.

You need to be perceived as a confident and competent authority who oozes certainty! Customers' brains will be comforted by this and will be drawn towards it. Customers' mirror neurons will start to register this feeling of comfort and certainty coming from you, and they will in turn experience it for themselves.

Any lack of comfort on your part will be registered, perhaps at an unconscious level, and this feeling of discomfort and uncertainty may lead to you losing the sale. For example, if your body language is inconsistent with your words, then your clients may consciously or unconsciously register this and decide that they can't trust you.

In a Harvard Business School working paper, Assistant Professor Amy Cuddy states that 'an examination of 185 videotaped two minute pitches showed that venture capitalists were far more likely to invest in entrepreneurs who displayed confidence, passion and enthusiasm' (Cuddy, Wilmuth and Carney, 2012). Customers have to buy *you* before they will buy your product or service!

Observing the customer

In order to become proficient at observing our customer's non-verbal communication we need to keep some key principles in mind:

- *Context.* Non-verbal communication needs to be observed in terms of the context the person is experiencing. If, for example, the customer rushes into the meeting with you very late, sweating profusely and out of breath from having run from the other side of the building to meet you then this would need to be taken into account! As you interact with your customers more frequently, you will start to establish what their default or baseline non-verbal communication is like, and you can then use this to provide a backdrop for any behaviour that is exhibited.

- *Clusters.* This chapter will describe a number of behaviours that could be indicators that the customer is feeling comfortable or uncomfortable. One behaviour or indicator on its own should not be taken as evidence that the customer is feeling one way or another. Attune yourself to look out for 'clusters' of behaviour that when considered together provide more robust evidence of how the customer is feeling.
- *Changes.* Pay close attention to any changes in non-verbal communication, as these are usually indicative of the internal state the customer is experiencing. The most immediate non-verbal indicator or display is usually the most accurate, as it has been generated from the reactive limbic system before the customer has a chance to exert any form of conscious control or influence over it.
- *Congruence.* Pay attention to any difference or incongruence between what is said verbally and what customers' body language is saying. Any differences need to be noted. If customers say, for example, that they are comfortable with your proposal, but are indicating high levels of non-verbal discomfort then this is a rumble strip moment! You need to slow down and probe more deeply by asking some further questions such as 'Is there any aspect of the proposal that you would like to be changed?' The incongruence is not necessarily a sign of lying or deceit (although it might be – more of that later!); it may just mean that the customer is feeling uncomfortable about some aspect of your proposal and feels uncomfortable about sharing this with you. The Blue customer, for example, prefers to avoid any conflict and may not vocalize some concerns with you for fear of causing any offence.

So with this in mind let us go on a 'top to toe' journey through your customer's non-verbal communication!

The head

When customers are feeling comfortable their head may tilt over to one side or the other. This can be a sign of friendliness, comfort or an attempt to build rapport and is also believed to be a submissive behaviour as well as one used during courtship!

During your sales pitch should customers' heads suddenly straighten upright then this could be an indicator that they have seen or heard something that they don't like.

The forehead

When someone is experiencing emotional or physical pain either directly or when empathizing with another person the central forehead contracts. For this reason it is sometimes called the 'grief muscle'. The forehead is a very good place to detect if the customer is feeling uncomfortable, as it presents real-time limbic information on how the customer is feeling.

The furrowed forehead is usually a good sign that the customer is feeling uncomfortable, anxious, concerned or confused. If the customer is feeling comfortable the forehead will be smoother and unfurrowed (see Figure 14.1).

FIGURE 14.1 Furrowed forehead

The eyebrows

Lowering of the eyebrows is an indicator that the customer may disagree with you or have doubts or uncertainties. And if the customer raises one eyebrow this is a possible sign of scepticism.

Raised or arched eyebrows generally indicate comfort, confidence and positive feelings. This can also indicate surprise or disbelief.

The eyes

They have been described as the mirrors to our soul, perhaps because the eye's retina is an outgrowth of the forebrain, so looking into someone's eyes could be rather like looking into the brain itself.

The eyes take in a huge amount of information and pass it to the brain, but they can also provide a useful insight into what is going on inside the brain. For example, the pupils of the eyes dilate (enlarge) or constrict (narrow) in response to light conditions. However, they also respond to emotional changes. When we see things we are excited by, we would like or that could be rewarding, our pupils expand more than they would usually. And if we see something that we dislike or find uncomfortable they contract more than they would usually. These changes are unconscious, involuntary and beyond control and therefore are good indicators of the customer's level of comfort or discomfort.

In his book *Manwatching*, ethologist Desmond Morris (1978) mentions that jade dealers in pre-Revolutionary China deliberately wore dark glasses in order to conceal their excited pupil dilations when they were handed a particularly valuable specimen of jade. Before they did this, their pupil dilations were consciously watched for by jade salesmen. When they saw interest they put their prices up! Perhaps there is something for us to learn from these salespeople of old!

When customers feel comfortable with what they are seeing, the eyes and the muscles around them will be relaxed and soft. When customers feel uncomfortable the eyes and the muscles around them will be harder and squinting may occur. Squinting or a narrowing of the eyes is a sign of discomfort or dislike.

Squinting is one form of what is known as eye blocking. Other forms include customers covering or shielding their eyes with their hands and lowering their eyelids for prolonged periods (see Figure 14.2). This is an instinctive and hard-wired behaviour if we see something we don't like or that makes us uncomfortable.

FIGURE 14.2 Eye blocking

You can also tell the sincerity of customers' smiles by whether their eyes are engaged in the process or not. When a smile is genuine and sincere (what is called the zygomatic or heartfelt smile) the corners of the customer's mouth will come upwards towards the eyes and they become involved in the smile, with the outer corner of the eye crinkling into 'crow's feet'. If the smile is less genuine or false then the smile does not involve the eyes and no crow's feet will be seen.

When we are emotionally aroused or stressed (such as when excited, when lying and during courtship) our eyelids blink faster than the normal blink rate of about 15 to 20 closures per minute and can even increase to over 100 closures per minute.

The nose

If customers are feeling uncomfortable and perhaps even stressed they will touch and massage their nose more often (see Figure 14.3). Despite what you may have read or heard, touching the nose is not on its own a reliable indicator of lying. It is more likely to be an indicator of anxiety and discomfort, which may possibly be as a result of the increased tension people experience when lying.

FIGURE 14.3 Nose rub

When customers are excited about something you may notice that their nasal wings dilate. This appears to be a response to the brain needing more oxygen. This can also be a sign that customers want to take some action.

The mouth

If customers are feeling comfortable their lips will be relaxed and loose. If the customer feels uncomfortable or anxious you will notice that their lips tighten up, press together and almost seem to disappear (see Figure 14.4).

Possible signs of discomfort or stress include licking of the lips, rubbing the tongue backwards and forwards along and across the lips, lip biting and mouth touching.

The chin

If customers touch or stroke the very end of their chin this is often a sign of thought and consideration and should be taken as a positive indicator that they are considering your proposal carefully (see Figure 14.5).

FIGURE 14.4 Disappearing lips

FIGURE 14.5 Chin consideration pose

The neck

If customers are feeling stressed or uncomfortable they may stroke or ventilate (by pulling at the neck of their shirt or top) the neck area. When customers' hands go to their neck there is usually something they are uncomfortable about. They may also rub or scratch the neck behind their ear if they are feeling uncertain about something (see Figure 14.6).

FIGURE 14.6 Neck scratch or rub

The arms and hands

One of the most commonly misunderstood non-verbal communication displays is crossing the arms. Contrary to popular opinion this is not necessarily a sign of negativity or defensiveness. To some people crossing their arms is a comfortable or comforting thing to do. If, however, the arm fold is tight or they appear to be almost gripping themselves, then it is likely to be an indicator of discomfort.

It can also be a blocking behaviour, which is an unconscious defence against something customers find uncomfortable. They may also block you with a folder or bag held up to their chest (see Figure 14.7). This is a sure sign that they are uncomfortable.

If customers are wringing their hands together or have interlaced their fingers and are rubbing or stroking them together then this is a good indication that they are feeling uncomfortable about something.

However, if customers rub the palms together then this indicates that they are anticipating a positive result.

FIGURE 14.7 Blocking

A sign of frustration or of holding something back is that customers will clench their hands together either in front of themselves on their desk or in their lap. If you see this gesture then it is a good time to ask if they have any questions or concerns.

An important point to note when involved in the sales process or the negotiation that invariably forms part of it is to be aware that open palm gestures are usually associated with honesty and openness. They are saying 'I have nothing to hide.' If you speak with your palms downwards then this will make you appear more commanding and can be a useful behaviour to use during negotiations to help you to get your point across.

The chest

If customers are feeling comfortable with you and your proposal they will unconsciously turn their chest towards you. And if they are feeling uncomfortable then they will turn their chest away from you. The extreme of this behaviour is if customers turn their back on you. Not a good sign!

As an aside, research by ex-FBI body language expert Joe Navarro (2009) shows that people will perceive you to be more open and honest if they can see your torso. So turning towards people and unbuttoning your jacket to allow what Navarro calls a 'ventral display' will help customers to feel more comfortable with you.

The hips

If you see customers shifting their weight from side to side on their hips or shifting around in their chair (even subtly) then this is a sign that they are uncomfortable with what is being discussed or shown.

The legs

If customers' legs are crossed in a loose and/or low position on the legs then this is usually a sign of comfort (see Figure 14.8).

FIGURE 14.8 Low leg cross

Crossing of the legs (when either sitting or standing) is often an indicator of comfort. The act of crossing the legs takes one or both feet off the floor. If customers are feeling uncomfortable, unconsciously the limbic system will get them to place both feet flat on the floor – so they can make a fast escape if needed!

A possible sign of discomfort is customers crossing their legs and placing the ankle of one leg on to the knee of the other leg in a high leg cross, effectively turning the legs into a kind of barrier between them and you (see Figure 14.9).

FIGURE 14.9 High leg cross

If customers' legs are loose and splayed this is a comfort signal; if they are together and tight this is a sign of discomfort.

The feet

If customers' feet (and legs) are pointing towards you then this is a strong unconscious indicator that they are feeling comfortable with you and what you are saying. If, however, their feet are pointing away from you then they are not comfortable and unconsciously their feet are pointing in the direction they wish to go – away from you (see Figure 14.10).

FIGURE 14.10 Foot point-away

So there we go – a journey through your customers' non-verbal behaviour from top to toe!

Keep your eyes open and observe customers' non-verbal communication. Look out for signs of comfort that mean they are feeling positive about you and your proposal. Their body is exhibiting 'towards reward' signs. And if you see a cluster of displays of discomfort behaviour then this needs to act as a non-verbal rumble strip telling you that you need to slow down (and perhaps even reverse!), as something is making the customers feel uncomfortable. So, when you see it, slow down and ask the customers a question to check how they are feeling. Maybe they have a query or concern that you can answer, which will make them feel comfortable once more before proceeding with the sale.

In Chapter 15 we are going to explore how to steer a sale to a conclusion by negotiating effectively.

15 Neuro-negotiating

Although this is predominantly a book about selling it would be remiss of me not to include some content on the subject of negotiation. Selling and negotiating are two essential skills that you have to master if you want to prosper as a sales professional. Selling and negotiating are both inextricably linked together in the sales process. Although the boundaries between them are blurred, they are two distinct stages and two different skill sets.

Selling is about convincing customers to purchase a product or service, or to enter into some form of arrangement or agreement with you. Selling can be defined as establishing a need or want to buy (remembering that people tend to buy what they want rather than what they need), and then matching the benefits of your product or service to that need or want. These benefits and how they help customers to get what they want are articulated in your sales proposal or value proposition.

Negotiating on the other hand is about agreeing the terms upon which the purchase, arrangement or agreement will take place. This may include many factors, such as volume purchased, delivery schedule and method, purchase frequency, amount of payment, timing of payments, service levels, product or service configuration and so forth.

To maximize your profit margin the golden rule is: sell first; negotiate second. The reason for this sequence is that the more convinced customers are of the benefits of your product or service the more they are likely to be prepared to pay for it. Selling is about communicating the value of what you have to offer. The more value customers perceive a product or service to give them, the higher the price they will be prepared to pay for it.

On some occasions selling alone may be enough. You may be able to convince customers to purchase your product or services without any negotiation taking place. However, in the majority of modern selling situations, you will be drawn into negotiation.

If you allow yourself to be drawn into negotiation too early (and experienced buyers will attempt to do exactly this), you are weakening your negotiating power and missing out on the opportunity to convince customers of the benefits (and therefore the value) that your product or service will bring them. To prevent this happening it is important to focus on three distinct stages in the sales process. These are in sequence: planning and preparing; selling; and *then* negotiating.

In the course of my work as a speaker and consultant working with countless sales professionals, the situation that I encounter with most of them is as illustrated in Figure 15.1. A small amount of planning is conducted. In my experience far too few salespeople plan and prepare well enough for customer meetings, and as a result of this the depth and quality of their selling are limited. They will then frequently find themselves dragged far too quickly into the negotiation stage (which lessens their ability to communicate value) by customers, who are deliberately trying to tip the balance of power in the negotiation in their favour.

FIGURE 15.1 An all-too-common scenario

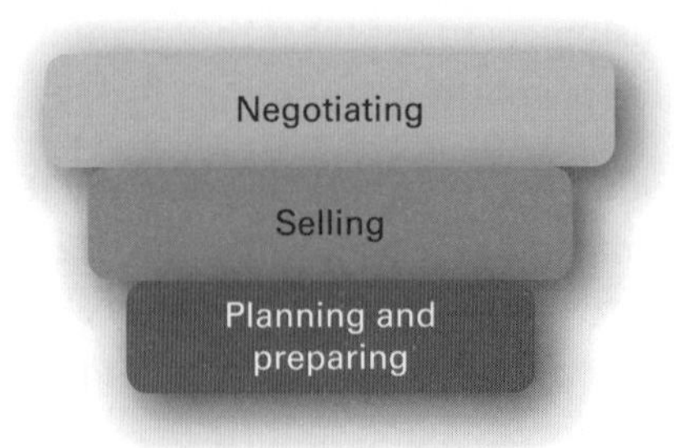

The ideal scenario is as illustrated in Figure 15.2, where salespeople plan and prepare for both the selling and the negotiating stage thoroughly. They then enter the selling stage, spending sufficient time to understand customers' needs and wants and then articulating a powerful value proposition. They then make the transition into the negotiating stage, maximizing profit margins by conducting an effective negotiation that is built on a firm foundation of planning, preparing and good-quality selling.

FIGURE 15.2 The ideal scenario

If the foundation of planning and preparation is weak, the selling stage will usually be too shallow and short and will lead to the power balance in the negotiation being tipped in the favour of customers. The customers will then exploit their advantage, and salespeople's profit margins inevitably suffer as a consequence.

We covered the importance of planning in Chapter 8, and if you take the time to build a solid foundation of planning and preparation for your customer meetings then your ability to sell will improve, which will in turn lead to you feeling stronger and more confident in the negotiation stage. The more confident you are feeling, the better your profit margin is likely to be.

Why (most) salespeople aren't good at negotiating

In my experience, in the majority of cases salespeople just aren't as good at negotiating as they need to be. Why is this?

At an early stage in their career salespeople are usually told to 'keep the customer happy'. They have been taught that happy customers are good customers, and they will go out of their way to placate unhappy customers. Customers know this and will deliberately make salespeople uncomfortable by appearing to be 'unhappy' as a way of tipping the power balance in their favour. They are using the principle of 'stay away from pain' and 'move towards reward' with the salespeople. They create an uncomfortable situation for salespeople, which the salespeople are motivated to move away from and provide a way to move towards a more comfortable (and therefore rewarding) situation. And all the salespeople have to do to feel comfortable is to give the buyer a much lower price!

Customers (particularly professional procurement buyers, whom salespeople are encountering more and more often) will attempt to shortcut the sales stage and pull the salespeople out of their selling comfort zone and into their negotiating discomfort zone. Experienced buyers will increase the levels of discomfort as much as possible using psychological ploys and tactics. Uncomfortable salespeople will often pay their way out of discomfort in the form of (at best) some form of financial concession or (at worst) a non-reciprocated 'give-away'. They will literally pay their way out of discomfort with their employer's profit margin. For salespeople to become more effective negotiators they need to recognize this and 'feel comfortable feeling uncomfortable'!

Two distinct skill sets

Selling and negotiating are two distinct skill sets. Although selling and negotiating are inextricably linked there are distinct differences. In selling we are attempting to persuade, convince, enthuse, justify and explain. By contrast in negotiation we are stating our position, considering, making and weighing proposals and making demands for what we want.

In the majority of cases salespeople are far less comfortable with negotiating than buyers are. If salespeople receive sales training, the majority of the time will be spent on the process of selling and very much less time (if any at all) will be spent on the process of negotiating. Professional buyers (more and more salespeople are encountering procurement professionals) on the other hand will usually receive training only in negotiation. Therefore when it comes to the negotiation stage of the sales process buyers usually have the advantage.

Having spent countless hours running realistic and demanding negotiation simulations, I have often seen salespeople's selling comfort zone once again rearing its ugly head in terms of negotiation behaviour. Salespeople will tend to do far too much information giving in the negotiation, driven by their predilection for persuading and selling. In doing so, they miss out on gathering the necessary information that would enable them to make effective negotiation proposals. They spend far too much time thinking about things from their perspective and not enough time where their focus should be – getting inside the customer's brain and understanding things from the customer's perspective. On the other hand professional buyers will tend to exhibit higher levels of information gathering, giving them far more knowledge and information that they can use to their advantage. In negotiation, knowledge is power.

To negotiate successfully, you need to understand – really understand – what the other person wants to achieve. When you fully understand this you can create a deal that meets the needs of both sides. If you fail to gather enough information (as salespeople frequently do), your chances of securing a profitable deal are very limited. Sales professionals need to move their own agenda and objectives from the front of their mind to the back of their mind. The customer's needs and priorities need to be in the front of the salesperson's mind – and they usually are not.

The importance of feeling comfortable feeling uncomfortable

With the powerful motivating forces of 'stay away from pain' and 'towards reward' being leveraged by the customer on us in negotiation situations, it is important that firstly we are aware of the power of these forces on our behaviour. Having read this book you will be acutely aware of the power of these forces, and indeed will become adept at using them to your advantage in selling situations.

Secondly, we must take steps to neutralize their impact upon us and find ways to use them to tip the negotiation in our favour. Taking the time to mentally 'step back' and take a few deep breaths will enable our cortex to have greater control and influence. We need to calm our own 'gremlin brain', allowing our more rational cortex to run the show. In doing so we can calm and tame the powerful hard-wired impulses to stay away from pain and move towards reward.

When we begin to feel understandably anxious during negotiations, remember that our 100,000-year-old brain has detected what it instinctively regards as a potential threat and is equipping us to deal with it. You feel nervous, your mouth feels dry, your limbs shake, you start to sweat, you start to breathe rapidly and you want to go to the toilet! You are feeling nervous and scared. Or are you?

When you are faced with what your primitive brain views as a challenge your emotional reactions send your mind and body through a series of chemical changes to help you deal with the situation. This involves your brain, nervous system and adrenal glands. Your heart rate, muscles, and energy and concentration levels are brought to their most effective to help you to cope with the situation. When we are faced with a real or perceived threat, the more primitive parts of our brain trigger the 'freeze, fight or flight' response. This is your built-in survival instinct being activated. This response occurs not only in genuinely dangerous situations but also in situations that we perceive to be threatening. Despite the fact that we are living in a sophisticated and civilized society we are still using our 100,000-year-old brains. That is, we still react to perceived threats in the same way our Neanderthal ancestors did!

We carry these patterns of behaviour with us from prehistoric times, as an effective survival mechanism, and it works automatically without us having to think about it. This automatic response was highly useful if you were about to be attacked by a prehistoric creature, but it has its disadvantages in modern business negotiations. We have been programmed over millions of years of evolution to be able to instantly summon extra speed and strength when faced with a threat. When faced with danger, or what we perceive as danger, the body and mind go almost instantly through a series of complex chemical changes to help us to deal with the situation effectively. The adrenal gland releases a large amount of adrenalin into the blood. Our breathing and heart rate quicken. Our blood pressure rises, and sugar is released into the blood to give extra energy. Our muscles increase activity, and our lungs work faster. Our pupils dilate to give better vision, and our metabolism is accelerated by our thyroid hormones. Some of our physical systems, such as the digestive tract, are shut down to allow more blood to move to the muscles. The intestinal digestion of food slows down, our bladder sphincter muscle contracts, our sweat pores open and our saliva becomes thick and viscous.

Perhaps you can see how these physical reactions are the same physical reactions we might associate with being scared, nervous or anxious. The dry mouth, the need to go to the toilet, perspiring, feeling nauseous and breathing heavily are commonly associated with nervousness, fear and feeling scared. In fact you are not really 'scared' or 'nervous' at all; it is the effect of hormones such as cortisol, adrenalin, noradrenaline and dopamine. Your emotional reactions have triggered these chemical changes to help you to cope with the threat, be it real or imagined. Your brain is trying to be helpful!

So, when you feel as though you are nervous going into a negotiation, you are in fact incorrectly interpreting the physical signs. These physical signs are in fact positive; they are your brain and body 'tooling up' to perform well under stress. Your body and mind want to bring your heart rate, muscles, energy and concentration to their most effective levels.

Get used to this feeling; indeed welcome it, as when harnessed positively it will give you a performance edge. Recognize it is happening, harness the feeling and do not misinterpret it for something you have to move away from. Instead embrace it, feel comfortable feeling uncomfortable and you will be in a strong state of mind to negotiate from.

As a professional speaker, I am very used to stepping on to a stage in front of a large audience of up to a thousand people. Every single time I do this, I experience the chemical changes described above. I welcome them and indeed thrive on them, as I believe this gives me an 'edge' when I am on stage. And I experience the same feelings before an important sales pitch or negotiation. I am grateful that my brain is doing its best to support me in doing a good job! I also believe that it means that I am taking the pitch or negotiation seriously and not getting arrogant or complacent. I want to be fully 'switched on' when delivering a speech or a sales pitch or conducting a negotiation, as it means that I perform to a far higher standard.

The five stages of negotiation

If you want to enhance your capability and confidence as a negotiator, it is vital to understand that most negotiations go through five distinct stages. These five stages are:

- *Step 1: planning and preparing.* Far too many negotiators fail to plan and prepare correctly! This vitally important step is often seriously overlooked, as the negotiators are too keen to get stuck into the action! Effective planning and preparation are the hallmark of the professional negotiator. If you do not plan and prepare properly, you can only react to what happens in the negotiation rather than leading and controlling it.

- *Step 2: discussing and/or arguing*. Depending upon the subject and the people involved, this stage can be a relatively calm discussion or a raging argument – or something in between. Whatever the nature of the conversation taking place, the purpose of this stage of the negotiation is to review the issue(s) and to exchange information. It is good practice to make every possible effort to understand the other party's point of view and to make sure that the other party understands yours.
- *Step 3: signalling and proposing*. Each negotiation that you become involved in will have two possible solutions – one that meets all of your needs and one that meets all of the other party's needs. In reality the final agreement usually falls somewhere in between these two ideal solutions. Therefore as a negotiator you need to be on the lookout for signals or signs of willingness from the other party to consider movement. Signals are usually followed by proposals. A proposal is a suggested action, approach or process that one party in a negotiation makes to the other party. Proposals advance negotiations. Without them not a lot happens!
- *Step 4: bargaining*. This stage of the negotiation is characterized by the two parties trading with each other. Variable items are traded so that both parties can achieve their objectives. The key to effective bargaining is giving to get. Never make a concession without getting something of equal or greater value in return. No freebies!
- *Step 5: closing and agreeing*. Closing is when agreement to proceed is reached. At this stage you have a deal!

The importance of planning and preparation

The most important stage in the entire negotiation process is step 1: planning and preparing. Up to 90 per cent of your success as a negotiator is related to the quality of the planning and preparation that you do in advance of the negotiation.

Here are some of the vital elements that professional negotiators will include in their planning and preparation:

- *Objectives.* What specifically do you want to achieve and how will you measure your success? Make sure you write your objectives down, as this will make them more concrete. Unclear objectives will usually lead to poor results. It is also vitally important to consider the objectives that the other party may have and then check these are correct during the discussion stage. Get out of your head and into the other party's head!
- *Negotiation parameters.* As most negotiations will fall somewhere in between the ideal outcome of both parties, it is important to consider the range within which a deal is possible. Define your ideal outcome (your 'like'), a realistic outcome based upon your knowledge to date (your 'intend') and finally your walk-away point (your 'must'). Then consider the likely range of the other party. If you cannot secure your 'must' then you must walk away. Not all deals are worth doing. For example, it is a good idea to think carefully in advance of the absolute minimum price you will be prepared to accept, a realistic price that provides you with a reasonable margin and an ambitious price that would deliver a very good margin. You can then open your price negotiation above your 'like' figure and negotiate from there.
- *Negotiable areas.* What are the elements that the negotiation will be orientated around, for example contract length, specification and so on? List areas that are important to you and those that you anticipate are important to the other party. During the negotiation you will attempt to get some of what you want by trading something the other party wants in return. Work out what each concession will cost you so that you can make sure that you always get something of equal or greater value in return.

Although we could go into the subject of negotiation planning and preparation in much greater depth, if you commit to always considering these vital areas then your confidence will increase, and this will enable you to negotiate far more profitable deals.

The four different negotiators

In Chapters 6 and 7 we met the four different 'colours' or behavioural preferences that we will encounter when meeting our customers. We looked in some detail at how to identify them and then adapt our behaviour when selling to these different behavioural preferences.

Now let me add to this with some considerations to keep in mind when negotiating with, as distinct from selling to, these four different behavioural preferences.

The Green negotiator

Green negotiators will be more outgoing, social and spontaneous. Make additional efforts to maintain this positive relationship with them during the negotiation, as the tension that sometimes accompanies negotiations may stimulate their fears of rejection. They dislike confrontational approaches to negotiation, as they would prefer to look for creative solutions and ideas rather than slugging it out!

As they like choices and options they will be open to look at new and creative solutions. This sort of behaviour will be stifled if they are feeling tense: hence the need to put additional positive focus on the relationship. Remain positive and upbeat during the negotiation.

The Blue negotiator

Blue negotiators will be more calm, open and warm. Make additional efforts to build on your existing relationship, as this will be important to them. They are the behavioural preference that is most uncomfortable with conflict or confrontation and will avoid it if at all possible. Match their calm nature.

They will probably want to hear your opinion or proposal first and will want to consider it fully before responding. Be open and honest in your views; they will need to trust you.

They are patient and steady negotiators and will take their time. They will often want to consult with others to gather their views before bringing the negotiation to a close.

They do not like radical change or pressure, so do not push them to make a commitment. Negotiations with the Blue preference will take longer, and it may take a number of meetings before the deal can be closed.

The Red negotiator

Red negotiators will be tough in their approach and will want, and expect, to take charge of the negotiation, set the pace and be in control all the way through. Negotiations with Reds can be fast and furious!

If you attempt to 'fight' them for control then they are likely to respond negatively. Paradoxically, you can exert greater 'control' over Red negotiators by allowing them the sense that they are in the driving seat. Use suggestions and subtle hints to nudge them in the direction you want them to go in. Offer them options that they can choose from, as this helps them to feel in control, and let them know that you are willing to explore different approaches to achieving a result.

They may appear rude, arrogant and impatient, but this is due to their high focus on task, achievement and results. Help them to negotiate a deal that lets them have a sense of 'winning' by achieving the results they need and they will participate more fully.

The Gold negotiator

Gold negotiators can be challenging to deal with because they negotiate using facts, data and proof. You need to be very well planned and prepared (they will be!) and ready to answer a lot of detailed questions.

They want to negotiate the right deal and make the right decision the first time, so they want to be certain, and they will attempt to do this by gathering concrete proof and evidence. You will need to be very patient. Any attempt to push things forward will be resisted.

They may very well want to discuss the fine detail of the 'small print', and once their questions are answered and they have the information they need they will start to feel comfortable with moving forward with a proposal in the negotiation.

They can appear cold and unemotional as negotiators, but this is just the impact of their preferred behavioural style being amplified by the demands of the negotiation.

Different negotiation styles

Good negotiators are able to flex their negotiation style as required. For example, if you are making a one-off transactional purchase such as a car, a house, a new kitchen or double glazing where you will not be having an ongoing relationship with the other person, then you may choose to adopt a tough, hard-bargaining approach. This is sometimes referred to as a 'win/lose' style. This style is characterized by power plays, intimidation and hard-nosed tactics.

The downside of this style is that, as the more primitive part of the brain is very active here, trust and flexibility are low. In this style I only benefit if you lose. I get more if you get less. You can see how this situation will stimulate the reptilian and emotional brain, which is why these styles of negotiation can be very combative.

For negotiations that involve longer-term relationships a different style is required, a style that looks to deliver a result for both parties, to find, create and add value to the deal, and the relationship is often referred to as a 'win/win' style. To succeed with this style you need to keep your negotiations brain-friendly, limiting the perceived threat response and keeping the reptilian and emotional brain in check as much as is possible.

Although in reality the subject of negotiation style is more complex than a simple division between win/win and win/lose styles, this broad definition will provide a useful working model for us to use.

The power/comfort balance

Power, or rather the perception of power, has a huge bearing on negotiation success. Invariably the people who perceive that they are in the more powerful position will get the better deal. Even with a 'win/win' approach it does not mean that everything is divided equally or fairly. It is a win/win result if both parties leave the negotiation feeling comfortable or at least with an acceptable deal. The challenge is to conclude a win/win deal with the balance of the win tipped in your favour! Therefore it is important that you are communicating your power effectively so that the other party perceives you as being the more powerful party. Much of this perception will be unconscious and therefore will be based upon what the reptilian and emotional brains are exposed to.

On the other hand we also need to continue to keep the customer's brain feeling as comfortable as possible. The negotiation stage is likely to be potentially the most uncomfortable or threatening stage for the customer's brain.

If we are not careful we can all too easily trigger a threat response in customers' brains, with the resulting reaction, which may hamper our ability to negotiate successfully with them. You will recall that if customers' reptilian and emotional brains become over-stimulated this will prevent the prefrontal cortex from functioning fully, interfering with a more rational approach to negotiation and decision making.

If this happens customers will have more difficulty accessing long-term memory, they will find it more difficult to remember what has been said, their ability to judge will diminish, their ability to solve problems will be impaired, they will be more sensitive to any perceived threats and they are more likely to react negatively.

Therefore we will be looking to achieve a delicate balance between radiating power and confidence and taking actions to maximize comfort for the customer's brain.

Comfort builders

To build comfort, continue to adapt your behavioural style to best suit that of the customer. The notes above on how to negotiate with each of the four behavioural styles need to be incorporated into what you have learned about them already.

Avoid any language, behaviour or actions that could irritate or stimulate the customer's reptilian or emotional brain. Things to avoid include:

- interrupting;
- talking over the other person;
- not listening to what the other person is saying;
- being confrontational:
  - 'Take it or leave it!'
- scoring ego points:
  - 'So I'm right then?'

- being dismissive:
  - 'That'll never work!'
  - 'That's a stupid idea!'
- being sarcastic:
  - 'Wow, I bet that took a lot of doing!'
- being provocative:
  - 'That is a very generous offer!'
  - 'What you fail to realize is...';
- using dismissive language:
  - 'That's no good to us!'
  - 'I hear what you are saying but...';
- using phrases repetitively:
  - 'With respect...';
- finger pointing;
- gesturing quickly and excessively;
- raising your voice;
- excessive staring;
- moving too close to the other person and invading personal space or 'territory';
- becoming emotional (think of the impact on the customer's mirror neurons!).

If one or more of the above behaviours are displayed by one side in the negotiation then there is always the danger that the other side will counter back. When humans are psychologically 'pushed' they will often 'push' back. It only takes three or so of these psychological pushes before we are in an attack-and-defend-style loop, with the behaviour of one negotiator provoking the other negotiator, who in turn provokes back and away we go. The reptilian and limbic systems respond to the threat response, and the negotiation rapidly spirals downwards as the cortex loses its somewhat delicate hold over the proceedings. It takes human beings about three or so pushes before the situation degenerates into a squabble or fight. If this happens, call a timeout so both sides can calm down, take a break, take some deep breaths and get the cortex back in control.

A better way is to avoid such instances occurring by focusing on comfort-building behaviours. Things you can do to increase comfort during the negotiation include:

- listening fully and attentively to what the other person says;
- using positive and encouraging language:
  - 'I would be happy to explore...'
  - 'Let us discuss how we could move this forward'
  - 'My intention is to fully understand your position on this'
  - 'I am committed to doing all I can to make this positive for both of us';
- making steady but comfortable eye contact;
- smiling when appropriate;
- sitting in a relaxed, upright and open body posture, uncrossing your arms and legs;
- nodding when the other person makes a point that you agree with;
- rewarding flexibility on the other person's part:
  - 'Thank you for that consideration';
- building on the other person's suggestions:
  - 'That's a good idea and we could add to that by...';
- maintaining an even and calm voice modulation;
- speaking at a steady pace.

What follow are some power principles that may be used on you or that you may choose to use to build the customer's perception of your power.

Power builders

One principle that can be very effective is what is called 'power posing'. This is based upon research carried out by Amy Cuddy, an assistant professor at Harvard Business School (Carney, Cuddy and Yap, 2010; Cuddy, Wilmuth and Carney, 2012). As described in Chapter 14 our non-verbal communication reflects how we are feeling. If people are feeling powerful they will often adopt what Cuddy called 'high-power poses' – open, expansive postures and gestures. For example, you will see the hands-and-arms-outstretched victory gesture when sportspeople triumph during a sporting event, or powerful businesspeople may lean back in their chair, put their arms behind their head and put their feet up on their desk (see Figure 15.3). When people feel powerful they become bigger by expanding the space their body occupies by wide-spaced limbs and spreading out. This reflects and communicates the power they feel.

FIGURE 15.3 High-power pose 1

One very powerful non-verbal display of power is to stand with your hands on your hips with your elbows out (see Figure 15.4). When combined with a wide stance (the more space between our legs the stronger we appear), this is a very commanding and territorial display. You will see this posture adopted by people in leadership and authority positions such as army drill sergeants and police officers. A variation that can be seen displayed by members of the British royal family is where the arms are placed behind the back and the hands are held together.

By contrast when people are feeling less confident and powerful they use 'low-power poses', which are more closed, contractive postures, folding their arms, hunching their shoulders, making themselves appear smaller by collapsing the body inwards (see Figures 15.5 and 15.6).

FIGURE 15.4 High-power pose 2

FIGURE 15.5 Low-power pose 1

FIGURE 15.6 Low-power pose 2

This behaviour is not confined to the human animal. Across the animal kingdom alpha male primates bulge their chest, the male peacock spreads his colourful tail or the submissive dog shrinks down towards the floor in deference to the more powerful dog. From a hormonal point of view alpha male animals have higher levels of testosterone (what Cuddy describes as 'the dominance hormone') and lower levels of cortisol (the 'stress hormone'). When an animal first occupies alpha status its testosterone levels rise and its cortisol levels fall. If it loses its alpha status its testosterone levels fall and its cortisol levels rise.

So it would appear that leaders tend to have relatively high testosterone levels and relatively low cortisol levels. These would be the calm and confident people who can go into a potentially stressful situation and not react adversely to the stress.

A calm and confident state strikes the ideal balance between comfort and power. It is a calm, authoritative power in oneself that is not likely to provoke an adverse reaction in the customer's reptilian and emotional brain.

Cuddy and her team noted that power poses were displayed when people were feeling powerful as a reflection of that feeling, but were fascinated to know whether adopting the power poses in advance helped people to feel powerful. In a series of fascinating experiments the researchers took saliva samples from test subjects (both men and women) to measure their testosterone and cortisol levels. Some test subjects were asked to adopt 'high-power poses' and some were asked to adopt 'low-power poses'. Each test subject adopted the relevant poses for two minutes. After some additional exercises a second saliva sample was taken and the results compared with the first. High-power poses caused an increase in testosterone of about 19 per cent and a decrease in cortisol levels of about 25 per cent. Low-power poses caused a decrease in testosterone and an increase in cortisol. In short, the high-power poses made people feel more powerful, and their hormone levels proved it.

So let us look at how to apply this and several other power principles:

- Adopt a power pose for at least two minutes before going into negotiations. Perhaps the easiest one to adopt is the hands on hips posture illustrated in Figure 15.4. This can also be adopted whilst waiting in the customer's foyer or lobby prior to the meeting.
- Do not sit down whilst waiting to be collected from the customer's foyer or lobby or waiting for the negotiation to commence. Stay standing up in your power pose. Standing up also keeps your energy levels higher and means that when the customer meets you for the first time you are at eye level.
- In the animal kingdom the alpha male tends to sit or lie on higher ground than the more subordinate members of the pack. Standing up when greeted means that you are not 'lower' than the customer when you first meet. Also watch out for the old buyer's trick of putting salespeople into a low chair so that the buyer towers over them!
- Some people may try to 'power handshake' you by turning the palm of their hand downwards when offering you their hand. The only way to initially take their hand is by turning your hand palm upwards into what is believed to be a submissive hand gesture. Should this be done to you, the counter is to initially take the hand and then on the first downward handshake turn the other person's hand back sharply to the more normal vertical position.

- Control the other person's time. Keeping people waiting is a classic power play. The message is clear – 'I will see you when I am ready to see you because I am in charge.' If this is done to you either occupy yourself with an important task (a phone call, for example) while you are waiting and then keep the other person waiting while you finish off or call the person's bluff and say that you have another meeting to attend and offer to reschedule.
- Control the other person – make the appointment on a day and at a time that suits your diary and not any date option the other person offers (even if it suits you). When showing the person into the meeting room issue a series of seemingly polite requests in order to get the person to comply with you, for example 'Please take a seat' (and gesture to a specific seat), 'Make yourself comfortable – take off your jacket' or 'Please pass me the water.' All of these actions are starting to exert subtle control over the other person.
- When in the negotiation continue to sit in an open and expansive manner. Drape your arm over the chair next to you, or stand and lean an arm on the flip chart or whiteboard if you are using one. Lean back slightly in your chair and use a 'steepling' gesture. Steepling involves touching the spread fingertips of both hands together without the palms touching (see Figure 15.7). It is called steepling because the shape the hand makes resembles a church steeple. Steepling signifies that you are feeling confident. It shows you have confidence in and are comfortable with what you are saying. It has been observed that high-status people in society often steeple.
- Slow down your body movements and gestures. Appear unhurried and in control.
- Use a steady voice speed and tone. If you use more downward voice inflections your voice will appear more commanding. An upward voice inflection indicates a question. A downward voice inflection indicates a command.
- As described in Chapter 14, open palm gestures are indicative of openness and honesty, and gestures made with the palms downward are perceived to be more commanding and domineering. So when making statements or proposals during negotiation there may be times that a downward palm gesture will help to drive your point home. You can combine the downward palm gesture with a downwards voice inflection for maximum impact!

FIGURE 15.7 Steepling

- Apply time pressure. Usually the person with the most urgency to do the deal (for example, the salesperson who needs to hit this month's sales target) has the least power. If you suspect that the other person has a deadline you could propose slowing things down to see if there is a reaction.
- Watch out for good cop/bad cop. One negotiator plays the role of good cop and is nice, pleasant, welcoming and friendly. Another negotiator plays the role of bad cop and is mean, unpleasant, demanding and hostile. This plays on the hard-wired 'stay away from pain' and 'towards reward' circuits. They will bounce you backwards and forwards between good cop and bad cop. After a while good cop will make you an offer that seems attractive because it seems preferable to another encounter with bad cop. In reality they are working together to play you! Recognize it, and perhaps even ask them: 'Hey, come on, are you guys playing good cop/bad cop?'

By monitoring the power/comfort balance carefully (by using the 'stay away' and 'towards reward' non-verbal indicators discussed in Chapter 14) you can make adjustments as you go through to negotiate a good deal.

Is the customer lying?

When I am speaking on the subject of negotiation, or running seminars on the subject, I am often asked if it is possible to spot verbal or non-verbal signs that indicate that the person you are negotiating with is lying.

Despite what you may have heard, the truth of the matter is that foolproof non-verbal indicators of lying have not yet been discovered. In addition, we need to draw a distinction between outright deceit and being careful and considered about what information we reveal and when during a negotiation. Although the more open you are able to be, the easier it is to build trust in a negotiation, there are times when it is wise to play one's cards close to one's chest in a negotiation and consider when to reveal certain pieces of information. Therefore it is possible to misinterpret prudent commercial caution as deceit.

That being said, here are some possible indicators that might provide you with an indication that the other person is not telling the whole truth:

- *Voice pitch.* People who are lying will speak with a higher voice pitch than truth tellers.
- *Voice speed.* Liars will speak more slowly than truth tellers.
- *Duration of message.* Liars' answers to questions will be shorter than truth tellers' and there will be a lack of detail. They will tend to give general rather than specific answers.
- *Language.* Liars' language will seem more negative than truth tellers'.
- *Self/other references.* Liars use fewer references to themselves and others than truth tellers.
- *Response time.* Liars take longer to respond to a question than truth tellers.
- *Hesitation.* Liars hesitate when speaking more than truth tellers.
- *Speech errors.* Liars' speech contains more errors and is less fluent than truth tellers'.
- *Non-verbal incongruence.* Liars leak non-verbal incongruence, such as making a micro-nod when making a negative statement or a micro-head shake when making a positive statement.
- *Eye contact.* Liars blink more often and have more dilated pupils than truth tellers. Many people believe that a lack of eye contact is

a sign of deceit. This is very well known, and liars will go to extra lengths to make steady eye contact.

- *Gestures.* Liars tend to gesture less, touch less and move their arms and legs less than truth tellers.

Please remember, as mentioned in Chapter 14 on non-verbal communication, to keep your eyes and ears open for clusters of the behaviours described above. One indicator is unlikely to mean that the other person is being untruthful. If you pick up a cluster of them, however, then you might want to take extra care!

To close this chapter here are a few brain-friendly negotiation tips that you can take and apply:

- Set your objectives, be confident and aim on the high side. Work out your 'like', 'intend' and 'must' and anticipate those of the other party.
- Be firm and flexible. Be persistent in pursuing your objectives but not rigid in pursuing any particular solution. Good negotiators are flexible negotiators. There may be a way to give them what they need but package it in a way that works for you.
- Listen more than you speak. Give them a good listening to! Being listened to is comforting, as it shows you are paying attention to them.
- Ask lots of direct questions and listen carefully to the answers. Gather more information than you give, although be as open and honest as is commercially prudent to increase trust and comfort levels.
- Focus on the needs that underlie the positions people take in the negotiation. Go looking for needs, ask questions about why they are important and then see if you can structure the deal to meet both party's needs.
- Summarize on a regular basis to clarify what has been discussed to give a sense of progress and to subtly exert your power over the negotiation. The person who summarizes more often usually has more control over the negotiation.
- If things get heated or confusing call a break. Take time out. Get some fresh air, take some deep breaths and give your brain a break.

- Give nothing away free. Always give to get. Get something of equal or greater value for any concession you make. Make your proposals conditional, using the 'If you... then I...' format: 'If you give me the five years' free warranty and servicing and the metallic paint then I will buy the car.' The person has to give what you have asked for to get the concession.
- Trade things that are low-cost to you but valued highly by the other party and vice versa. What has the other party got that won't cost that party much but would be very valuable to you?
- Establish the principle that 'Nothing is agreed until everything is agreed.' It is only once every element of the deal has been agreed that you will agree. This prevents other parties 'cherry-picking' the elements they like and avoiding the areas they don't like.

Once you have reached agreement, confirm everything verbally and check for understanding. Then confirm everything in writing and check that both parties agree with the written agreement. Then you have made a deal! Now all you have to do is to make sure that you and the other party deliver on the deal.

Executing what you said you would do and doing it rapidly and professionally will continue to build the customer's confidence in you and comfort with you. Over time, as you continue to build the customer's confidence and comfort levels, you begin to build a 'defensive wall' of positive 'towards reward' feeling towards you around the customer's business. The positive feelings that the customer gets from working with you will help to protect your customer from attack from your competitors.

As any sales professional knows, selling to and trying to steal a happy, satisfied, well-serviced customer is one of the toughest sales jobs around, so make life as difficult as you can for your competitors by making sure your customers feel very comfortable and confident in their relationship with you.

16 Conclusion

So here we are at the end of the book! I hope you have enjoyed reading *Neuro-Sell*. However, what is far more important than the reading of this book is for you to apply what you have learned. It is only through action and the application of the powerful principles outlined in this book that you will improve your sales performance.

Please treat this book as a practical tool to be used. Take one area of your sales process that you think needs the most improvement and start making changes there first. A chain is only as strong as its weakest link. It is only through the application and practice of what you have learned in this book that you will be able to make significant positive changes in your sales performance.

Most of this book has been focused on your customer's brain. As we conclude, let us spend a little time considering your own brain. Your brain is constantly changing, rewiring and making new connections between the billions of neurons it contains. This is called 'neuroplasticity', which refers to changes in neural pathways and synapses. These occur as a result of changes in your environment but more importantly through changes in your behaviour and neural processes, for example your thinking, learning and actions. You can choose to allow neuroplasticity to happen by chance, or you can take a choice to direct the changes consciously. Dr Jeffrey M Schwartz, a psychiatrist at the UCLA School of Medicine, calls this 'self-directed neuroplasticity' (Schwartz and Begley, 2003). When we are changing our brains on purpose we are employing self-directed neuroplasticity.

So my challenge to you, over the next few days, weeks and months, is to make a consistent and concerted effort to apply the principles in this book.

Keep this book with you, refer to it, apply what you have learned and you will see the difference in your sales performance. Engage in self-directed neuroplasticity to develop and fine-tune your brain to be a highly effective sales instrument by continued application of the principles and practices in this book.

With the knowledge you now possess about *PRISM* Brain Mapping, study your profile (if you have not already done so please visit **www.neuro-sell.com** to download your free *PRISM* Brain Mapping profile) and consider the stronger preferences you have and how you can leverage these more fully. Consider areas that are lower preferences for you and ensure you are aware of these and manage them carefully. In addition to your free *PRISM* Brain Mapping profile, **www.neuro-sell.com** gives you access to a variety of other resources and downloads you can use to maximize your sales performance with neuroscience.

I would love to hear about your successes. Nothing gives me greater pleasure than hearing from fellow sales professionals about how they have applied what they have learned and the success this has brought them.

If you are interested in booking me to speak at your conference or event, or would like me to provide consultancy services and/or train your sales leadership or sales force, or if you are interested in licensing my 'brain-friendly selling' process for your organization to use then I look forward to speaking with you. I promise to give you a good listening to!

You can contact me directly at **simon@simonhazeldine.com**. If I can be of any assistance to you on your journey to sales mastery then please do get in touch.

Good luck and good neuro-selling!

Simon Hazeldine
www.simonhazeldine.com
www.neuro-sell.com
Twitter: @simonhazeldine

References

Bargh, JA, Chen, M and Burrows, L (1996) Automaticity of social behaviour: direct effects of trait constructs and stereotype activity on action, *Journal of Personality and Social Psychology*, **71**, pp 230–44

BBC (2013) *Horizon: The creative brain: how insight works*, Television programme, BBC London, 14 March

Blakeslee, S (2006) Cells that read minds, *New York Times*, 10 January

Carney, DR, Cuddy, AJC and Yap, AJ (2010) Power posing: brief nonverbal displays affect neuroendocrine levels and risk tolerance, *Psychological Science*, **21** (10), pp 1363–68

Cialdini, RB (1993) *Influence: Science and practice*, HarperCollins College Publishers, New York

Condon, WS and Ogston, WD (1966) Sound-film analysis of normal and pathological behaviour patterns, *Journal of Nervous and Mental Disease*, **143**, pp 338–47

Cowan, N (2001) The magical number 4 in short term memory: a reconsideration of mental storage capacity, *Behavioural and Brain Sciences*, **24** (1), February, pp 87–114

Cuddy, AJC, Wilmuth, CA and Carney, DR (2012) Preparatory power posing affects performance outcomes in social evaluations, Working paper 13-027, Harvard Business School

Dijksterhuis, A *et al* (1998) Seeing one thing and doing another: contrast effects in automatic behaviour, *Journal of Personality and Social Psychology*, **75**, pp 862–71

Dunbar, R (1998*) Grooming, Gossip and the Evolution of Language*, Harvard University Press, Cambridge, MA

Gobet, F and Clarkson, G (2004) Chunks in expert memory: evidence for the magical number four... or is it two?, *Memory*, **12** (6), November, 732–47

Goleman, D (1989) Brain's design emerges as a key to emotions, *New York Times*, 15 August

Gordon, E (2000) *Integrative Neuroscience: Bringing together biological, psychological, and clinical models of the human brain*, Harwood Academic Publishers, Singapore

Hall, ET (1998) *The Hidden Dimension*, Bantam Doubleday Dell, New York

Hazeldine, S (2011a) *Bare Knuckle Negotiating: Knockout negotiation tactics they don't teach you in business school*, Bookshaker, Great Yarmouth

Hazeldine, S (2011b) *Bare Knuckle Selling: Knockout sales tactics they don't teach you in business school*, Bookshaker, Great Yarmouth

Hazeldine, S (2012) *The Inner Winner: Performance psychology tactics that give you an unfair advantage*, Bookshaker, Great Yarmouth

Hazeldine, S and Norton, C (2012) *Bare Knuckle Customer Service: How to deliver a knockout customer experience every time*, Bookshaker, Great Yarmouth

Iyengar, SS and Lepper, MR (2000) When choice is demotivating: can one desire too much of a good thing? *Journal of Personality and Social Psychology*, **79** (6), pp 995–1006

Kahneman, D, Slovic, P and Tversky, A (1982) *Judgement under Uncertainty: Heuristics and biases*, Cambridge University Press, Cambridge

Kendon, A (1970) Movement coordination in social interaction, *Acta Psychologica*, **32**, pp 100–25

Knight, S (2008) The heart of selling, *Financial Times*, 13 September

Lakhani, D (2005) *Persuasion: The art of getting what you want*, Wiley, Hoboken, NJ

Lehrer, J (2009) *The Decisive Moment: How the brain makes up its mind*, Canongate, Edinburgh

Macrae, CN and Johnston, L (1998) Help, I need somebody: automatic action and inaction, *Social Cognition*, **16**, pp 400–17

Maxham, JG, III (1997) The role of adaptive selling in sales training: a salesperson perspective, in *Advances in Marketing*, ed JA Young, DL Varble and FW Gilbert, pp 195–203, Southwestern Marketing Association, Terre Haute, IN

Morris, D (1978) *Manwatching*, Grafton, London

Navarro, J (2009) *The Power of Body Language*, Nightingale Conant, Wheeling, IL

Nunes, JC and Dreze, X (2006) The endowed progress effect: how artificial advancement increases effort, *Journal of Consumer Research*, **32**, pp 504–12

PBS (2005) *NOVA scienceNOW*, Television programme, Season 1, Programme 1, 25 January

Pradeep, AK (2010) *The Buying Brain*, Wiley, Hoboken, NJ

Sanitioso, R, Kunda, Z and Fong, GT (1990) Motivated recruitment of autobiographical memories, *Journal of Personality and Social Psychology*, **59**, pp 229–41

Schumpeter, J (1950) *Capitalism, Socialism and Democracy*, Harper, New York

Schwartz, JM and Begley, S (2003) *The Mind and the Brain: Neuroplasticity and the power of mental force*, Regan Books, Los Angeles

Szegedy-Maszak, M (2005) Your unconscious is making your everyday decisions, *US and New World Report*, http://health.usnews.com/usnews/health/articles/050228/28think_2.htm

Than, K (2005) Scientists say everyone can read minds, *LiveScience*, 27 April

Zaltman, G (2003) *How Customers Think: Essential insights into the mind of the market*, Harvard Business School Press, Boston, MA

Further reading

Carter, R (1998) *Mapping the Mind*, Orion, London

Collett, P (2003) *The Book of Tells*, Doubleday, London

Condon, WS and Ogston, WD (1966) Sound-film analysis of normal and pathological behaviour patterns, *Journal of Nervous and Mental Disease*, **143**, pp 338–47

Cutmore, TRH *et al* (1997) Imagery in human classical conditioning, *Psychological Bulletin*, **122**, pp 89–103

Damasio, A (2006) *Descartes' Error: Emotion, reason and the human brain*, Vintage, London

Dijksterhuis, A *et al* (1998) Seeing one thing and doing another: contrast effects in automatic behaviour, *Journal of Personality and Social Psychology*, **75**, pp 862–71

Dunbar, RIM, Duncan, NDC and Marriott, A (1997) Human conversational behavior, *Human Nature*, 8, 231–46

Fine, C (2006) *A Mind of Its Own: How your brain distorts and deceives*, Icon Books, Cambridge

Gengler, CE, Howard, Daniel J and Zolner, K (1995) A personal construct analysis of adaptive selling and sales experience, *Psychology and Marketing*, **12**, July, pp 287–304

Goleman, D (1996) *Emotional Intelligence: Why it can matter more than IQ*, Bloomsbury, London

Gordon, E *et al* (2008) An 'integrative neuroscience' platform: application to profiles of negativity and positivity bias, *Journal of Integrative Neuroscience*, 7 (3), pp 345–66

Green, M and Brock, TC (2000) The role of transportation in the persuasiveness of public narratives, *Journal of Personality and Social Psychology*, **79**, pp 701–21

Harish, S and Weitz, B (1986) The effects of level and type of effort on salesperson performance, Working paper, Pennsylvania State University

Heath, C and Heath, D (2008) *Made to Stick*, Random House, London

Hsu, J (2008) The secrets of storytelling: why we love a good yarn, *Scientific American*, September

Iacoboni, M (2009) *Mirroring People: The science of empathy and how we connect with others*, Picador, New York

Kendon, A (1970) Movement coordination in social interaction, *Acta Psychologica*, **32**, pp 100–25

Klaff, O (2011) *Pitch Anything*, McGraw-Hill, New York

Knowles, PA, Grove, SJ and Keck, K (1994) Signal detection theory and sales effectiveness, *Journal of Personal Selling and Sales Management*, **14**, Spring, pp 1–14

LeDoux, J (1999) *The Emotional Brain: The mysterious underpinnings of emotional life*, Phoenix, London

Lehrer, J (2009) *The Decisive Moment: How the brain makes up its mind*, Canongate, Edinburgh

Levy, M and Sharma, A (1994) Adaptive selling: the role of gender, age, sales experience and education, *Journal of Business Research*, **31**, pp 39–47

Lindstrom, M (2009) *Buyology*, Random House, London

McKee, R (1999) *Story: Substance, structure, style and the principles of screenwriting*, Methuen, York

Miller, GA (1956) The magical number seven, plus or minus two: some limits on our capacity for processing information, *Psychological Review*, **63**, pp 81–97

Navarro, J (2008) *What Every Body Is Saying: An ex-FBI agent's guide to speed-reading people*, HarperCollins, London

Navarro, J (2009) The body language of the eyes: the eyes reveal what the heart conceals, *Psychology Today*, December

Nierenberg, GI and Calero, HH (2001) *How to Read a Person like a Book*, Metro Books, New York

Pease, A and Pease, B (2006) *The Definitive Book of Body Language*, Orion, London

Pinker, S (1997) *How the Mind Works*, Penguin, London

Schwartz, B (2004) *The Paradox of Choice: Why more is less*, HarperCollins, London

Spiro, RL and Weitz, BA (1990) Adaptive selling: conceptualization, measurement, and nomological validity, *Journal of Marketing Research*, **27**, February, pp 61–69

Stephens, GA, Silbert, LJ and Hasson, U (2010) Speaker–listener neural coupling underlies successful communication, *Proceedings of the National Academy of Sciences of the United States of America*, **107** (32), pp 24–30

Tobias, RB (1993) *20 Master Plots and How to Build Them*, Writers Digest Books, Cincinnati, OH

Weinschenk, SM (2009) *Neuro Web Design: What makes them click?*, New Riders, Berkeley, CA

Wilson, TD (2002) *Strangers to Ourselves: Discovering the adaptive unconscious*, Harvard University Press, Cambridge, MA

Woodside, AG and Wilson, EJ (2000) Constructing thick descriptions of marketers' and buyers' decision processes in business-to-business relationships, *Journal of Business and Industrial Marketing*, **15** (5), pp 354–69

Index

NB: page numbers in *italic* indicate figures or tables

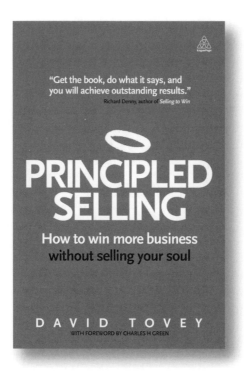

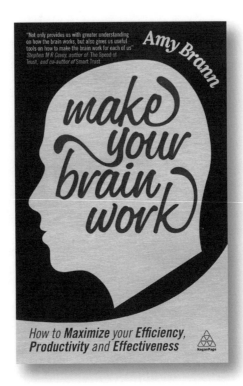